Praise for

Flying on One Engine:
The Bloomberg Book of Master Market Economists

EDITED BY Thomas R. Keene
BLOOMBERG NEWS

"An intriguing collection of well-written essays on the global economy from some of the best and brightest economic thinkers on Wall Street. **A relevant read and a good tool for anyone seeking to connect the day-to-day battles in the business world with the bigger picture**."

RICK WAGONER
Chairman and CEO, General Motors Corp.

"It is as if I entered an applied economics laboratory. This is **an excellent compilation of model-free insights about the relevant economic issues that are facing us—or may face us soon**. In addition, it has a wealth of facts and data from which the reader can build his own critical view. I came out convinced that there was real economics after all."

NASSIM NICHOLAS TALEB
Dean's Professor in the Sciences of Uncertainty
University of Massachusetts, Amherst
Author, *Fooled by Randomness*

"As one particularly concerned for some time about long-term fiscal responsibility, and the sustainability—or should I say unsustainability—of our entitlement programs and our current account deficits, I was particularly **pleased to see the very high quality of the commentators and their balanced approach in these crucial fields**. As for the other commentary, I learned a lot I did not know."

PETER G. PETERSON
Former Secretary of Commerce
Author, *Running on Empty*

"Turn to any essay in this innovative collection and you can see how Tom Keene has captured the unique niche that business economists fill in our economy. And because it is a 'real time' record, importantly covering the uncertain period following the 9/11 terrorist attacks, you can see who was right, who was wrong, and, more importantly, why. Lipsky and Glassman go out on a limb and predict the end of the jobless recovery. Malpass finds reasons for optimism following the gloomy data reported by Roach. Dudley and McKelvey draw a vicious circle and make the CBO look like Pollyanna. **Always provocative, practical, and timely, but rarely in agreement, these economists' writings are invaluable to policymakers.**"

JOHN B. TAYLOR
Undersecretary for International Affairs at the U.S. Treasury, 2001–2005
Raymond Professor of Economics, Stanford University
McCoy Senior Fellow, Hoover Institution

"All too often the insights generated by Wall Street economic thinking appear only in this week's newsletter, or that day's luncheon speech, and then vanish from view. Thomas Keene has gotten some of Wall Street's most prominent economists to step back and assess the global situation. **They don't all agree—that's what makes markets—but their often contrasting perspectives on the economic and financial challenges the world now faces are both interesting and instructive.** Many of these analyses will bear reading (and re-reading) long after today's market close."

BENJAMIN M. FRIEDMAN
William Joseph Maier Professor of Political Economy, Harvard University

"In this volume, sixteen of our most respected financial economists present their analysis of the current and imminent issues facing the global economy. **Any reader who delves into these pages is sure to broaden their appreciation of the key factors that are likely to shape the future for years to come.**"

MARTIN L. LEIBOWITZ
Managing Director, Morgan Stanley
Coauthor, *Inside the Yield Book*

Flying on One Engine

The Bloomberg Book of Master Market Economists

Flying on One Engine

FOURTEEN VIEWS ON THE WORLD ECONOMY

EDITED BY
Thomas R. Keene
BLOOMBERG NEWS

With a Foreword by KENNETH S. ROGOFF
and an Afterword by PETER L. BERNSTEIN

Viva Books Private Limited
New Delhi • Mumbai • Chennai • Kolkata • Bangalore • Hyderabad

First Indian Edition 2007

VIVA BOOKS PRIVATE LIMITED

- 4737/23 Ansari Road, Daryaganj, New Delhi 110 002
  E-mail: vivadelhi@vivagroupindia.net, Tel. 42242200
- Plot No. 76, Service Industries, Shirvane, Sector 1, Nerul, Navi Mumbai 400 706
  E-mail: vivamumbai@vivagroupindia.net, Tel. 27721273, 27721274
- Jamals Fazal Chambers, 26 Greams Road, Chennai 600 006
  E-mail: vivachennai@vivagroupindia.net, Tel. 28290304, 28294241
- B-103, Jindal Towers, 21/1A/3 Darga Road, Kolkata 700 017
  E-mail: vivakolkata@vivagroupindia.net, Tel. 22836381, 22816713
- 7, Sovereign Park Apartments, 56-58, K. R. Road, Basavanagudi, Bangalore 560 004
  E-mail: vivabangalore@vivagroupindia.net, Tel. 26607409, 26607410
- 101-102 Mughal Marc Apartments, 3-4-637 to 641, Narayanguda, Hyderabad 500 029
  E-mail: vivahyderabad@vivagroupindia.net, Tel. 27564481, 27564482

Published by arrangement with

Bloomberg L.P.
731 Lexington Avenue,
New York, NY 10022

ISBN 81-309-0532-9

Published by Vinod Vasishtha for Viva Books Private Limited, 4737/23 Ansari Road, Daryaganj, New Delhi 110 002.

Printed by Sanat Printers, Kundli - 131028.

Contents

Foreword

THIS FASCINATING COLLECTION of essays offers a window into a world most ordinary individuals have no access to: the insights and analyses of Wall Street's highly paid top economists. Sure, they are quoted in the press all the time, and you can see them on financial TV networks, but their most important analytical insights are featured in proprietary newsletters available only to a select few. These economists are not academics aiming to change how the world works twenty years from now, they are practical thinkers who think deeply about how complex financial markets work here and now. But don't imagine that they are naive about recent cutting-edge research. On the contrary, they are constantly on the lookout for new ideas about how the economy works, particularly new empirical research.

For students used to reading dry academic textbooks, the style of these essays is refreshing. First, most academics struggle to make their ideas relevant to policymakers and practitioners. The economists in this book are masters at achieving practical relevance; they understand the problems facing elite investors, and know how to laser in on practical ideas. Second, the writing style is enormously refreshing for students used to laboring through sometimes infuriatingly qualified (if this, this, and this, then that) academic research. On Wall Street, economists get paid to have a strong and unique point of view, and to be able to express it and defend it. Of course they are wrong sometimes. But for someone trying to understand how to grapple with real-world problems, it is enormously helpful to watch a pro who knows how to take a clear position and explain it.

And it is not all about making money. These essays contain important insights for global policymakers on problems ranging from dealing with the U.S. current account deficit to Social Security. Bringing together this stellar collection of Wall Street economists to write on important topics of the day can only serve to raise the level of economic discourse in public debate. This book may be the first of its kind. Let's hope it is not the last.

—KENNETH S. ROGOFF
Department of Economics
Harvard University

About the Contributors

Richard B. Berner is a managing director and chief U.S. economist at Morgan Stanley. He directs the firm's forecasting and analysis of the U.S. economy and financial markets.

Before joining Morgan Stanley in 1999, he was executive vice president and chief economist at Mellon Bank, and a member of Mellon's senior management committee. Previously, he served as a principal and senior economist for Morgan Stanley, as a director and senior economist for Salomon Brothers, as economist for Morgan Guaranty Trust Company, and as director of the Washington, D.C., office of Wharton Econometrics. He also served for seven years on the research staff of the Federal Reserve in Washington, where he co-directed the Fed's model-based forecast and was a member of the team that developed the Fed's first multicountry model used for international policy analysis. He has been an adjunct professor of economics at Carnegie Mellon University and at George Washington University.

Dr. Berner is a member of the economic advisory panel of the Federal Reserve Bank of New York, a member of the advisory committee of the Bureau of Economic Analysis, Department of Commerce, a member of the board of advisers of Macroeconomic Advisors, LLC, chair of the economic advisory panel of the Bond Market Association, and a member of the board of directors of the National Bureau of Economic Research. He is a past president and fellow of the National Association for Business Economics. He is the winner of forecasting awards from *Market News* and the National Association for Business

Economics, and has been a member of *Time*'s board of economists.

He received his bachelor's degree from Harvard College and his Ph.D. from the University of Pennsylvania. He researched his dissertation under SSRC-Ford Foundation grants at both the University of Louvain, Belgium, and at the University of Bologna, Italy.

Peter L. Bernstein is founder and president of Peter L. Bernstein, Inc., which he established in 1973 as economic consultants to institutional investors and corporations around the world. He is the author of *Against the Gods: The Remarkable Story of Risk* (Wiley, 1996), which has sold over 500,000 copies worldwide and has been awarded the Edwin G. Booz Prize and the Clarence Arthur Kelp/Elizur Wright Memorial Award from The American Risk and Insurance Association (ARIA). He is also the author of several other books including *Wedding of the Waters: The Erie Canal and the Making of a Great Nation* (W.W. Norton, 2005), *The Power of Gold: The History of an Obsession* (Wiley, 2000), and *Capital Ideas: The Improbable Origins of Modern Wall Street* (Free Press, 1992). He is the consulting editor of *The Journal of Portfolio Management* and his writings have appeared in professional journals such as *The Harvard Business Review* and the *Financial Analysts Journal,* and in the popular press, including the *New York Times,* the *Wall Street Journal,* and *Worth.* He lectures widely throughout the United States and abroad on risk management, asset allocation, portfolio strategy, and market history.

Mr. Bernstein has received three major awards from CFA Institute (formerly the Association for Investment Management & Research), the organization that administers the Chartered Financial Analyst program: The Award for Professional Excellence, CFA Institute's highest award; The Graham & Dodd Award, given annually for the outstanding article in the *Financial Analysts Journal* for the previous year; and The James R. Vertin Award, recognizing individuals who have produced a body of research notable for its relevance and enduring value to investment professionals. He graduated from Harvard College with a degree in economics, magna cum laude.

William C. Dudley is chief U.S. economist at Goldman, Sachs & Co. He is responsible for the economic and interest rate forecasts for the United States and Canada and is co-chair of the Retirement Committee, which oversees the management of the employees' pension fund and 401(k) assets. Dr. Dudley has briefed the Federal Reserve Board on several occasions and is a member of the technical consultants board to the Congressional Budget Office and the economics advisory panel to the Federal Reserve Bank of New York. He and has team are highly ranked by *Institutional Investor* (including being ranked at number one in 2000 and 2003).

Prior to joining Goldman Sachs in 1986, Dr. Dudley was vice president in charge of regulatory analysis at J.P. Morgan. Bill began his career as an economist in the Financial Studies Section at the Federal Reserve Board in Washington, D.C., where he occasionally briefed the Board of Governors on payments and bank regulatory issues.

He received his Ph.D. from the University of California at Berkeley in 1982.

Thomas D. Gallagher is a senior managing director and head of International Strategy & Investment's policy research team in Washington. He has been an *Institutional Investor*-ranked Washington analyst for each of the past twelve years, and was ranked number one in the last two years. Prior to joining ISI in 1999, Mr. Gallagher spent thirteen years at Lehman Brothers and eight years in the federal government. He graduated from the University of South Dakota in 1976, received his master's from Harvard's Kennedy School of Government in 1978, and is a chartered financial analyst.

James E. Glassman is a senior economist and managing director with JPMorgan Chase & Co. He works closely with the firm's global treasury and capital markets groups, and he is an adviser for the firm's key corporate relationships. He is the coauthor of *Global Issues,* a publication that explores a wide range of global economic and market issues. Dr. Glassman's views are widely cited in the financial media, where he is a frequent commentator on economic policy issues.

From 1979 through 1988, Dr. Glassman served in the Research and Statistics and Monetary Affairs Divisions at the Federal Reserve Board in Washington, D.C. His responsibilities ranged from coverage of inflation, to labor market developments, to money and capital markets trends, and finally to the Federal Reserve's monetary policy operating procedures. He joined Morgan Guaranty in 1988 and Chemical Bank in 1993, which, through mergers with Chase Manhattan, JPMorgan, and most recently Bank One, is now JPMorgan Chase & Co.

Dr. Glassman earned a bachelor's degree in economics from the University of Illinois and a Ph.D. in economics from Northwestern University.

David P. Goldman is a managing director and head of debt research for Banc of America Securities. In this role, he is responsible for supervising research in fixed income, foreign exchange, emerging markets, and commodities. He is based in New York.

Before joining the bank in June 2002, Mr. Goldman spent four years at Credit Suisse First Boston as a global credit strategist. In 2001, *Institutional Investor* named him the strategy generalist for the publication's All-American Fixed Income team. From 1996 to 1998, he was a principal at SG Capital LLC, an investment management company specializing in mortgage derivatives. From 1993 to 1996, he headed fixed-income strategy at Bear, Stearns, where he was responsible for research across all fixed income asset classes, including corporates, Treasuries, mortgages, high-yield, and emerging markets. While under his leadership, the team was lauded as a leading fixed income team by *Barron's*.

As a consultant to Bear, Stearns in 1992, Mr. Goldman worked with top Russian Federation officials on sovereign debt issuance. Between 1988 and 1993, he was chief economist of the consulting firm Polyconomics, Inc.

He received his bachelor's degree in history from Columbia University, where he graduated cum laude. He also studied in the doctoral program at the London School of Economics, and earned a master's degree in music from City University of New York. Mr. Goldman is a member of the board of governors at Mannes College of Music in New York City.

Robert D. Hormats is vice chairman of Goldman Sachs (International) and managing director of Goldman, Sachs & Co. He joined Goldman Sachs in 1982.

Dr. Hormats served as assistant secretary of state for economic and business affairs from 1981 to 1982, ambassador and deputy U.S. trade representative from 1979 to 1981, and as senior deputy assistant secretary for economic and business affairs at the U.S. Department of State from 1977 to 1979. He served as a senior staff member for international economic affairs on the National Security Council from 1969 to 1977, during which time he was senior economic adviser to Dr. Henry

Kissinger, General Brent Scowcroft, and Dr. Zbigniew Brzezinski. Dr. Hormats was a recipient of the French Legion of Honor in 1982 and the Arthur Fleming Award in 1974.

Dr. Hormats has been a visiting lecturer at Princeton University and is a member of the board of visitors of the Fletcher School of Law and Diplomacy and the dean's council of the John F. Kennedy School of Government at Harvard University. He is also a member of the Council on Foreign Relations. Dr. Hormats is a board member of the Irvington Institute for Immunological Research and of Engelhard Hanovia, Inc. He is also a member of the international advisory board of Toyota, as well as the advisory boards of *Foreign Policy* and *International Economics* magazines. In 1993, he was appointed by President Clinton in to the board of The U.S.–Russia Investment Fund.

Dr. Hormats's publications include *Abraham Lincoln and the Global Economy; American Albatross: The Foreign Debt Dilemma;* and *Reforming the International Monetary System: From Roosevelt to Reagan.* Other publications include articles in *Foreign Affairs, Foreign Policy,* the *New York Times,* the *Washington Post,* the *Wall Street Journal, American Banker,* and the *Financial Times.*

Dr. Hormats received his bachelor's degree from Tufts University in 1965 with a concentration in economics and political science. In 1966, he received a master's, and, in 1970, a Ph.D. in international economics from the Fletcher School of Law and Diplomacy.

John P. Lipsky is the chief economist at JPMorgan Chase, responsible for the firm's worldwide economic and policy views. He also directs JPMorgan's flagship *World Financial Markets* and *Global Issues* publications.

Previously, Dr. Lipsky served as Chase Manhattan Bank's chief economist and director of research. He had served as chief economist of Salomon Brothers, Inc. from 1992 until 1997. From 1989 to 1992, he was based in London, where he directed Salomon Brothers European Research Group. He joined Salomon Brothers in 1984, following a decade at the International Monetary Fund.

His professional activities include serving on the board of directors of the National Bureau of Economic Research, the Economic Club of New York, the American Council on Germany, and the Japan Society. He also serves as a member of the advisory board of the Stanford Institute for Economic Policy Research and of the Economics Subcommittee of the Bond Market Association. He is a member of the Council on Foreign Relations.

A graduate of Wesleyan University, Dr. Lipsky earned a bachelor's degree in economics. Subsequently, he was awarded a master's and a Ph.D. in economics from Stanford University.

David R. Malpass is the chief economist at Bear Stearns & Co., Inc. He joined the firm in February 1993. He writes economic and financial studies and discusses financial market conditions with institutional investors. His duties include economic forecasts, Washington analysis, and global investment themes. He is a member of the Economic Club of New York and the Council on Foreign Relations, and sits on the board of the Council of the Americas.

In 2003 and 2004, investment institutions voted Mr. Malpass one of Wall Street's top five economists in the *Institutional Investor* survey. His 2001 analyses warned of the deflationary recession, while his 2002 and 2003 pieces highlighted the economic and equity recovery.

Between February 1984 and January 1993, Mr. Malpass held a series of economic appointments during the Reagan and Bush administrations, including six years with Secretary James Baker at the departments of Treasury and State. He was also Republican staff director of Congress's Joint Economic Committee and senior analyst for Taxes and Trade at the Senate Budget Committee.

In his government positions, Mr. Malpass worked on an array of economic, budgetary, and international issues, including the 1986 tax cut, several congressional budget resolutions, the Gramm-Rudman budget law, the savings and loan bailout, NAFTA, the Brady plan for developing country debt, and fast-track trade authority. He was a member of the government's Senior Executive Service and testified frequently before Congress.

From 1977 to 1983, Mr. Malpass worked in Portland, Oregon, as a contract administrator, a CPA with Arthur Andersen's consulting group, and a financial manager.

Mr. Malpass received a bachelor's degree in physics from Colorado College and an MBA from the University of Denver. He was a National Merit Scholar Finalist and a Boettcher Foundation Scholar. He studied international economics at Georgetown University's School of Foreign Service, and speaks Spanish, French, and Russian.

Thomas Mayer is a managing director and chief European economist at Deutsche Bank. He and his team provide economic and interest rate forecasts for European countries to Deutsche Bank's clients as well as to the trade and sales desks of the bank. Previously, Dr. Mayer worked for Goldman Sachs and Salomon Brothers. Before moving to the private sector, he held positions at the International Monetary Fund in Washington D.C., and at Institut für Weltwirtschaft in Kiel, Germany. He has published numerous articles on international and European economic issues in professional journals and commented on these issues in the media. He received a Ph.D. in economics from the University of Kiel and is a CFA charterholder.

Edward F. McKelvey is vice president and senior economist at Goldman Sachs, with more than twenty years of experience on Wall Street. He currently supports Goldman's fixed-income trading and sales operations, advises clients of the firm on the U.S. economic and financial outlook, manages the U.S. economic forecast, and focuses on federal budget issues. On occasion, he has briefed the Federal Reserve Board, the Congressional Budget Office, and the U.S. Treasury Department. Dr. McKelvey began his career on the faculty of Williams College and spent nearly eight years on the staff of the Federal Reserve Board before moving to Wall Street. At the Fed, he conducted and later supervised research on inflation, unemployment, productivity, and the U.S. capital markets. He received his Ph.D. in economics from Yale University in 1975.

Tim O'Neill is former executive vice president and chief economist of BMO Financial Group, a diversified financial services firm that provides retail and investment banking and wealth management services in Canada and the United States. The department he led played a key operational and strategic advisory role to the Bank of Montreal's senior management. Before joining Bank of Montreal in 1993, Dr. O'Neill was president of a nonprofit research organization and, prior to that, was a university professor for sixteen years. In his research, teaching, and consulting activities, Dr. O'Neill has focused extensively on the structure and performance of the North American economy. His recent work has examined many facets of globalization and their impact on the economy and on financial services.

Dr. O'Neill received his bachelor's degree (with Honors) at St. Francis Xavier University, his master's at the University of British

Columbia, and Ph.D. at Duke University. He was the first Canadian economist to be elected to the Board of Governors of the National Association for Business Economics (NABE). He was elected president of NABE in September 2002 and completed his term in October 2003.

Stephen S. Roach is managing director and chief economist of Morgan Stanley, a leading global financial services firm. In this role, he oversees the firm's highly regarded team of economists located in New York, London, Frankfurt, Paris, Tokyo, Hong Kong, and Singapore.

Dr. Roach has been widely recognized as one of Wall Street's most influential economists. His published research has covered a broad range of topics, with recent emphasis on globalization, the emergence of China, productivity, and the macro paybacks of information technology. He is widely quoted in the financial press and other media, and his work has appeared in academic journals, books, congressional testimony, and on the op-ed pages of the *Financial Times,* the *New York Times,* the *Washington Post,* and the *Wall Street Journal.*

Before joining Morgan Stanley in 1982, Dr. Roach was vice president for economic analysis for the Morgan Guaranty Trust Company in New York. He also served on the research staff of the Federal Reserve Board in Washington, D.C., from 1972 to 1979, where he supervised the preparation of the official Federal Reserve projections of the U.S. economy. Prior to that, he was a research fellow at the Brookings Institution in Washington, D.C.

He holds a Ph.D. in economics from New York University and a bachelor's degree in economics from the University of Wisconsin.

Kenneth S. Rogoff is the Thomas D. Cabot Professor of Public Policy and professor of economics at Harvard University. He was previously the Charles and Marie Robertson Professor of International Affairs at Princeton University and served as chief economist and director of research at the International Monetary Fund from August 2001 to September 2003. He is an elected member of the American Academy of Arts and Science as well as the Econometric Society, and a former Guggenheim Fellow. Dr. Rogoff is also a research associate at the National Bureau of Economics Research, a member of the Council on Foreign Relations, and a member of the Economic Advisory Panel of the Federal Reserve Bank of New York, and on the Advisory Committee of the Institute for International Economics. He has also served as director for the Center of International Development at Harvard.

Dr. Rogoff has published extensively on policy issues in international finance, including exchange rates, international debt issues, and international monetary policy, and is co-editor of *NBER Macroeconomics Annual.* Together with Maurice Obstfeld, he is co-author of the 1996 graduate text/treatise *Foundations of International Macroeconomics.* Dr. Rogoff received a B.A. from Yale University summa cum laude in 1975, and a Ph.D. in Economics from the Massachusetts Institute of Technology in 1980. He was awarded the life title of international grandmaster of chess by the World Chess Federation in 1978.

David A. Rosenberg is Merrill Lynch's chief economist for North America, responsible for formulating and communicating North America economic, interest-rate, and earnings outlook to clients.

Mr. Rosenberg joined Merrill Lynch in May 2000 as the Canadian chief economist and strategist. He began his investment career in 1983 at the Bank of Canada as an economist. Prior to joining Merrill Lynch, Mr. Rosenberg was the senior economist at Nesbitt Burns, as well as at the Bank of Nova Scotia.

In the 2002 Brendan Woods International Survey, Mr. Rosenberg ranked number one in economics. He has bachelor's and master's degrees in economics from the University of Toronto.

Michael R. Rosenberg is former managing director and senior strategist of Harbert Management Corporation's Global Macro Hedge Fund. Before joining Harbert, Dr. Rosenberg was global head of foreign-exchange research at Deutsche Bank (1999–2004) and head of international fixed income research at Merrill Lynch (1984–1999). Dr. Rosenberg's research team at Deutsche Bank was voted the world's number-one foreign exchange research team in *Institutional Investor* and *Euromoney* magazines' annual polls. Dr. Rosenberg also managed Prudential Insurance Company's global bond portfolio over the 1982–1984 period, and was a senior foreign exchange/money-market analyst at Citibank from 1977 to 1982. Dr. Rosenberg has written two books in the field of exchange rate forecasting, *Currency Forecasting: A Guide to Fundamental and Technical Models of Exchange Rate Determination* (Irwin/McGraw-Hill, 1996) and *Exchange Rate Determination* (McGraw-Hill,

2003). Dr. Rosenberg holds a bachelor's degree in accounting from the University at Albany, a master's in economics from Queens College, and a Ph.D. in economics from Penn State University.

John Ryding is the chief U.S. economist and a senior managing director in Bear, Stearns Fixed Income Research. At Bear, Stearns Mr. Ryding is responsible for analyzing and forecasting U.S. economic and financial market developments.

Before joining Bear, Stearns in March 1991, Mr. Ryding was the senior economist on the Federal Reserve Bank of New York's Open Market Desk. During his time at the New York Fed, he published a number of research articles on housing finance, the mortgage-backed securities market, monetary policy, and corporate finance. Mr. Ryding graduated from Sidney Sussex College, Cambridge University, England, in 1980, and began his career as a professional economist when he joined the Economics Division of the Bank of England. During his ten years at the Bank of England, he became head of the Bank of England's Economic Forecasting Group and then was made an economic adviser in the balance-of-payments area. During his time at the bank, he published articles on the demand for money, econometric and macroeconomic modeling, and foreign exchange intervention in various academic journals.

Mr. Ryding is a regular commentator on economics and financial markets in the financial press, on radio, and on TV, including regular appearances on Bloomberg TV and radio.

Kathleen Stephansen joined the Credit Suisse First Boston Corporation (CSFB) in October 2000 as the director of Global Economics. Her research responsibilities encompass analyses on global economic trends, and her marketing responsibilities entail promoting the Global Economics research product to the firm's global institutional client base, with regular visits to the firm's top one hundred accounts around the globe. She publishes her findings in the firm's weekly publications, appears regularly in print media, TV and radio, and is frequently invited to give speeches to clients and business associations.

Prior to joining CSFB, Stephansen was the chief international economist, senior U.S. economist, and senior vice president in the Fixed Income Division of the Donaldson, Lufkin and Jenrette Securities Corporation. As co-head of economic research, she oversaw a wide range of analyses covering both global and U.S. economics and related capital flows.

She holds an undergraduate degree in economics from the Université Catholique de Louvain (Belgium) and graduate degrees in economics from the University of New Hampshire and the London School of Economics, where she did her Ph.D. research. She was selected to the YWCA Academy of Women Achievers in 1995.

Acknowledgments

These chapters were conceived and written by sixteen economists in airline terminals, on airplane trays, in moments away from their many obligations to clients, colleagues, and families. Thank you for your dedication to this project. Economists are often perceived as prima donnas. There was not one complaint, not once an expression of attitude. It has been my privilege to work with you. Ken Rogoff likewise stole precious time away from his own dual book projects to write our foreword; so too, Peter Bernstein gave generously of his time and of his decades of perspective. Thank you, both.

This project, like so many at Bloomberg L.P., was begun over an unscheduled cup of coffee. Thank you to Bill Inman of Bloomberg Publishing for his inspiration and leadership, and to all of Bloomberg Press for their attention to detail. I must give special thanks to Barbara Diez Goldenberg for book design and to production editors JoAnne Kanaval and Mary Macher for their great care in seeing the book to completion.

My editor Jared Kieling deserves singular acknowledgement for his many improvements of my efforts. Less obvious, and deserved of my heartfelt thanks, was Jared's patience as he and I sought to bring together in one book the distinctive voices and characters of these original economic minds.

This book, like many of my other efforts at Bloomberg, would not be possible without the enthusiasm and grace of Betsy Perry. Her planning and execution of Bloomberg economic seminars and forums is second to none. Her introductions to many of these economists, and to a number of others as well, have been invaluable.

At Bloomberg News, real-time coverage and interpretation of economic events is a global, team effort. Thank you to all of the News Service, Television, and Radio. A thank you especially to John McCorry, Bill Ahearn, Bronwyn Curtis, Ted Merz, Michael McKee, Craig Torres, and Suzy Assaad for advice and perspective. This book followed from the success of the "Chart-of-the-Day" available on the Bloomberg Professional service. Thank you to Joe Winski, Patrick Chu, and my present editors, Cary O'Reilly, Robert Greene, and Mark McQuillan, for laying the groundwork for this book. One path from this project has led to the radio show I host, *Bloomberg on the Economy.* Thank you to Al Mayers of Bloomberg Radio for his show-origination and leadership, and to Cynthia Costas, Marty Schenker, Ed Caldwell, Mike Clancy, and Ken Kohn. The collegial and mutually reinforcing efforts of my colleagues at Bloomberg News have contributed to our having a stronger set of economists within these pages.

In electronic news, having the time and resources to compile a book is a luxury. In my case, the exercise of that luxury is founded upon the dedicated work of the worldwide Bloomberg L.P. sales force. Thank you to Lex Fenwick, chief executive officer of Bloomberg L.P., and to Max Linnington, for the foundation that allows for efforts like this one. David Tamburelli deserves a special thank-you. As well, thank you to Judith Czelusniak for her efforts in getting the plane off the ground.

It is understood that a company's chairman must provide strategic leadership. More rare is to see a chairman's broad vision combined with the quiet application of tactical leadership. Peter Grauer, chairman of Bloomberg L.P., has that dual strategic and tactical gift. His counsel directly improved this project.

An editor-at-large must, of course, get approval for such a book from the editor in chief. Matt Winker, the founder and editor in chief of Bloomberg News, provided immediate, enthusiastic, and continued support. His insistence on the linkage of economics, politics, and markets has afforded me a wide, fortunate mandate. Thank you.

My three children, sometimes bored of their dad's economic talk, have been forgiving of the time invested in this project. There is hope. One has studied the nuances of Joseph Stiglitz and Kenneth Rogoff, another threatens to take microeconomics, and still another is riveted by the issues that were discussed during the recent presidential election.

Economics, in the end, is the study of scarcity. This book, however, and all that led to and from it would not have been possible without my wife Heather's unlimited understanding and complete love.

Introduction

Clarity—that is the distinction. The messages and e-mails come by the thousands across my Bloomberg terminal: daily notes, weekly essays, and occasional papers from the leading market economists of the day. All are capably written, hundreds outstanding.

They are smart. Most express an opinion, but an opinion not wrapped in certitude but instead tempered by respect for risk and for the uncertainty of outcomes. Many of these authors write gracefully. The distinction that separates the sixteen economists in this book from others is—clarity. These market economists combine first-rate academics, measured opinion, and command of the King's English with a rare ability to explain the simple and complex clearly. Underneath their macroeconomics is real mathematical and microeconomic rigor.

In their daily work, market economists typically write of the moment to hurried colleagues and clients. My goal was to allow my contributors to expand in areas of their special expertise. Another aim was to create a book that forms a reliable bridge from the dryness of textbook theory to the real-world excitement of applied capital-at-risk economics. Still another desire was to provide business and investment professionals with the best in thought-provoking writing on market economics—work that can lead to answers and also to deeper questioning and further study.

I have not edited these fourteen chapters into one voice. Each chapter stands alone. The chapters, though complementary, are presented with the character and tone of each author's original manuscript. And yes, they collegially agree to disagree with each other sometimes on economic theory and its application.

We start with a generous foreword by Harvard economist Kenneth S. Rogoff. I asked Dr. Rogoff to give an academic economist's perspective of market economics. He has delivered more.

If economics is politics and politics is jobs, then we should lead with a chapter on employment. From there, we move across the global landscape, stop by foreign exchange and political economics, then address domestic issues of the United States. We close with the linkage of economics to the debt markets and two chapters that describe different outcomes for the economic future of Europe and Asia.

Finally, the esteemed author Peter L. Bernstein shows us the place that these economists' work occupies in an even larger sphere—that of society and its ceaseless striving to understand risk and uncertainty.

I asked Bloomberg Press to design a sturdy book to be used and abused. My fondest hope is that you read it with sharp pencil in hand, and that you underline, insert comments, and note down your beliefs and disagreements. These economists write with passion; we should read with equal passion. Many of the authors are famous and are required reading on op-ed pages worldwide. A few sit at the right or left hand of presidents and premiers in times of crisis. Others may be new names to you. Take a chance. Within are original thinkers and lucid writers who deserve a far wider notoriety.

A special word on the events and data cited in these chapters: Ordinarily, market economics has a limited shelf life, its output skimmed, read, and discarded by an expectations-driven investment world. Yet the art of seeing into markets is not static but dynamic, and the vision of these economists extends beyond the near term, as befits their long experience. These economists wrote their chapters in the summer of 2004. There is a special value to reading market economics in hindsight—a hindsight that makes a reader wiser and better prepared to address issues of the here and now. I have edited this work in 2005, but have every hope that it will remain fresh well into the second half of this decade.

You will find short introductions before each chapter. Here are a few suggested chapter sequences that will get you started towards meeting these master market economists.

If concerned about U.S. economics, perhaps try Dudley/McKelvey on the deficit, then Berner's study of corporate profits, then jump to

Lipsky/Glassman on employment, tackle David Rosenberg on inflation, then finish with the political economy of Gallagher.

For a global dichotomy, try the productive caution of Roach, then the optimism of Malpass. Rest. Then O'Neill on trade, Michael Rosenberg on foreign exchange, then Mayer and Hormats on Europe and Asia, respectively. Finish with Stephansen on the interdependence of fiscal economies.

Bored-stiff students should read Ryding on all-things Wicksellian, then David Rosenberg, then Goldman on the linkage of debt markets and economics. Switch majors. Then come back to economics, read Gallagher, take the summer off, then re-read Goldman.

The Chartered Financial Analyst, quantitative-finance set might enjoy Michael Rosenberg, then Goldman, then Ryding. Only then read Lipsky/Glassman to see if there is a job out there.

For those needing a fix of gloom, perhaps Roach, then Mayer, then Dudley/McKelvey, then Michael Rosenberg, then re-read Roach. Optimists might try Berner, then Ryding, then Hormats, then O'Neill, and finish with a flourish with, of course, Malpass.

A final word: This is a serious business. The political turmoil of the times, including the increasing polarization of the electorate witnessed in the United States in 2004, begs for a deeper economic understanding. Kenneth Rogoff has said that we are flying on one engine. It is my fondest hope that these pages, these economists, will provide knowledge and wisdom that will allow us all to land safely, on two wheels.

—THOMAS R. KEENE
Bloomberg News

Flying on One Engine

1

Reviving Employment Reflects Powerful Labor Market Transformations

JOHN P. LIPSKY
JAMES E. GLASSMAN

JOHN LIPSKY AND JAMES GLASSMAN are the Hammerstein and Rodgers of market economics. Lipsky, the chief economist for JPMorgan Chase, writes with passion on the global capitalistic experiment. He combines first-rate economics with the courage to reach, to grasp, and to analyze. Glassman, senior U.S. economist also at JPMorgan Chase, counters with Rodgers-like precision—careful analysis and an insistence on detail that melds perfectly into Lipsky's assertive thrust. Separate, they are very good; together they make economic music. Here, Lipsky and Glassman address the topic of the day, employment.

Employment prospects moved to the top of the domestic policy agenda during the second half of 2003 and have remained there since. Job market developments in 2004 and beyond will influence investor perceptions about U.S. economic performance. Job market developments also will influence Federal Reserve policy, as the labor market is a proxy for pressures on domestic resources and therefore for potential inflation risks. Moreover, labor market performance provides a key guide regarding the extent and importance of the structural changes that have occurred in the U.S. economy over the past decade.

Perceptions about the health of the nation's job market are strongly shaped by the pace of net hiring and by the unemployment rate. Net hiring indicates that economic growth is outpacing the increase in the economy's productive potential. By definition, output growth that falls short of the economy's underlying productivity growth rate can be met without adding new workers. Thus, to assure that new job market entrants can find work, the economy must grow at least as fast as potential real GDP growth, currently in the neighborhood of 4 percent per year. To find work for those currently unemployed requires growth faster than potential.

Net hiring trends exert a powerful influence on perceptions about the economy's overall strength. However, net hiring is not a prerequisite for increasing labor income. Net job growth therefore is not a straightforward guide to demand indicators such as consumer spending. Typically, changes in pay rates and in average hours worked are more significant in determining household income trends than is net hiring.

The quality of new jobs also shapes public perceptions about the health of the U.S. job market. As new hiring has accelerated, worries have shifted to the quality of new jobs and to labor compensation trends. Notwithstanding assertions by some Wall Street analysts that new jobs are mostly low quality and low paying—and that globalization is "gutting" the middle class—the facts indicate otherwise. According to the Bureau of Labor Statistics, most new jobs added over the past year—in construction, professional services, and in supervisory personnel—were in categories that tend to pay above the median wage. At the same time, job losses have been centered disproportionately in factory jobs that pay below the median wage.

More broadly, it is premature to draw any firm conclusion about the potential impact that globalization might be having on the character of the U.S. job market. Although the economy has been expanding since late 2001, net hiring only began in late 2003. In fact, current employment may be 3.5 million to 7 million jobs below what could be characterized as full employment, according to several independent perspectives. Furthermore, strong productivity growth remains the most promising path to improving job quality. Nonetheless, strong productivity gains may inhibit early cycle net hiring, even though they contribute in the long run to real labor compensation growth.

In addition to hiring and job quality or wage trends, the unemployment rate also shapes public perceptions about the health of the job market, and hence about the economy in general. However, shifts in the unemployment rate have become more difficult to interpret, as labor force participation appears to have become more sensitive to broad economic trends than was the case previously; that is, the unemployment rate more clearly reflects both supply and demand considerations than has been the case in the past.

Of course, a stable or rising unemployment rate could indicate a strengthening job market—if improving growth prospects encourage new job market entrants. Conversely, a falling unemployment rate potentially could signal a weakening economy if a sufficient number of discouraged workers left the job market. While these possibilities have been noted in the past, they seem to have become more relevant in recent years, reflecting the greater sensitivity of labor supply to labor market conditions.

Productivity Gains Mean More Growth Is Needed to Motivate New Hiring

Until recently, the public debate about the job market focused on the so-called "jobless recovery." That is, the key puzzle has been how output could have increased so rapidly without producing new jobs. The debate has focused on outsourcing—or "offshoring"—as a key variable limiting the strength of U.S. job growth. Other commentary has suggested that rising labor costs—especially benefit costs—have inhibited new hiring. The principal explanation probably lies elsewhere, however. In particular, total employment will not expand unless the economy grows faster than businesses are able to boost productivity. Thus, the "hurdle" rate for job growth—that is, the minimum rate of GDP growth needed to produce net job gains—will vary over time, depending on how successful companies are in improving their productivity.

In fact, the two popular explanations for earlier weak employment growth—offshoring and rising labor costs—can be ruled out as significant influences on the jobless recovery puzzle. Outsourcing does not explain the gap between domestic output and hiring because the output that is created by "offshored" jobs does not add to U.S. GDP—"domestic" output—but rather is classified as imports. Rising labor costs—such as those resulting from higher health benefit costs—also do not represent an obstacle to new hiring. Although benefit costs are rising rapidly, wage gains have been slowing.

As a result of slowing wage gains, overall labor compensation gains were moderating in 2003. In fact, labor compensation gains slowed even as the economy accelerated. In response, unit labor cost fell, implying that firms' incentive to increase their total employment was strengthening. Indeed, with corporate profit margins near record highs by late 2003, the market signal to businesses was that profits could be increased by expanding operations and adding staff.

Indeed, for almost a decade it has been obvious that U.S. productivity growth was accelerating, suggesting that firms were improving their efficiency more rapidly than previously. By now, it appears that productivity is growing at an annual rate of about 3 percent on trend, a rate that is about triple the pace considered typical from the mid-1970s through the mid-1990s. With the U.S. labor force growing on trend at about 1 percent per year—a rate that is expected to be maintained in the coming years—the U.S. economy's growth potential today is double what it was during the 1970s to 1990s. That is, the U.S. economy can grow output at about 4 percent annually, versus the roughly 2 percent rate in previous decades. In these circumstances, it shouldn't be surprising that faster output growth would be needed today to motivate new hiring, compared with the previous period.

Domestic Demand Pacing Productivity— Clearing the Path to a Jobs Recovery

Growth in new hiring depends on businesses' expansion. The basic building blocks of business expansion are threefold: demand growth, profit growth, and the ample availability of financing. By early 2004, all three elements were in place, and U.S. firms were confident that they could be sustained. As a result, it became highly likely that net job growth would accelerate, eventually attaining an above-trend pace (that is, sufficient to lower the unemployment rate).

That output has been expanding for some time—beginning within weeks following September 11, 2001—is deceptive. Although it is impressive that output grew at all during that time—especially given the broad-based consensus that the economy would suffer a serious recession in the wake of stock market losses and terrorist attacks—the expansion for nearly one and a half years didn't keep pace with the economy's growth potential. In response to this sustained sluggishness, U.S. policymakers enacted an unprecedented combination of budgetary and monetary stimulus.

The Federal Reserve eventually followed a path that had been common in previous downturns: reducing the federal funds rate to match that of the core inflation rate. In this case, the funds rate eventually fell to 1 percent, reflecting the Fed's goal of establishing a zero real funds rate in the context of the lowest inflation rate in more than forty years. In retrospect, there is little doubt that the actions by the Fed helped to buoy demand by underpinning asset values. However, the effects of Fed actions are felt most directly by the business sector, reflecting the sector's net debtor status.

While the Fed's response to the recession and the subsequent period of below-trend growth was typical in outline, the speed and scope of the fiscal reaction was unprecedented. In addition to the action of so-called "automatic stabilizers," discretionary tax cuts and spending increases together created a "fiscal thrust" equivalent to about 3–4 percent of GDP during 2001–2004—the most powerful discretionary fiscal boost of the post–World War II era. The starting point was a fiscal surplus equivalent to 3 percent of GDP in 2000, and the federal deficit reached a maximum of a bit more than 3.6 percent of GDP during 2004. By comparison, the deficit totaled more than 5 percent of GDP in the mid-1980s, following the early 1980s tax cuts, although the fiscal thrust at that time was modest.

The quick restoration of positive GDP growth by 2002—aided by expansionary policy—did not result immediately in net job gains. Clearly, companies don't hire new employees until they are confident that business is improving. Moreover, they will not hire new staff unless they can count on sales expanding faster than their ability to boost the productivity of their existing workforce. In other words, businesses will hire new staff only when that option offers the greatest profit potential. In general, this will occur only if growth in final demand outstrips the trend growth in productivity.

It was not until mid-2003 that GDP growth exceeded 3 percent on a year-on-year basis. As a result, it is not surprising that net job growth did not begin until then. With businesses increasingly confident in the sustainability of significant demand growth—as reflected in strong readings in purchasing managers' surveys in both the manufacturing and nonmanufacturing sectors—it is also not surprising that hiring accelerated in early 2004.

Thus, it was only by second half 2003 that the pre-conditions for net job gains were attained, with year-on-year growth in final demand outstripping the trend increase in productivity. Predictably, that is when net U.S. employment growth turned positive for the first time since 2000. However, in the hard-hit manufacturing sector, net job gains have not yet taken hold.

There is always a degree of uncertainty regarding which measures most accurately portray developments in the employment market. The two basic sources are the official payroll survey (a survey of employers that measures the total number of civilian jobs) and the household survey (that measures the total number of individuals who are employed). Typically, the household survey is more favorable than the count based on the survey of firms (see ***Figure 1.1***). In part, this pattern reflects the inclusion in the household survey of self-employed individuals. In addition, the payroll survey may not be capturing fully the new hiring at small businesses and startups.

With output growing by 4 percent in 2004 and likely to continue growing at an above-trend pace for the next several years, job totals will grow on a sustained basis even if productivity growth remains relatively rapid—as expected—and profit margins remain high.

"Productivity gains" are an elegant label for cost cutting. Thus, periods of unusually rapid productivity gains typically reflect either periods of notable technological advances or early cycle output gains following a slowdown, when firms have been hoarding resources in advance of improving demand. In any case, productivity gains tend to mirror financial pressures on firms to improve their profitability.

Productivity gains since the end of the recession in 2001 have been unexpectedly impressive, reaching an annual rate of 5–6 percent—

FIGURE 1.1 **Growth Drives Hiring**
Real GDP and Employment, Percentage Change from 12 Months Earlier

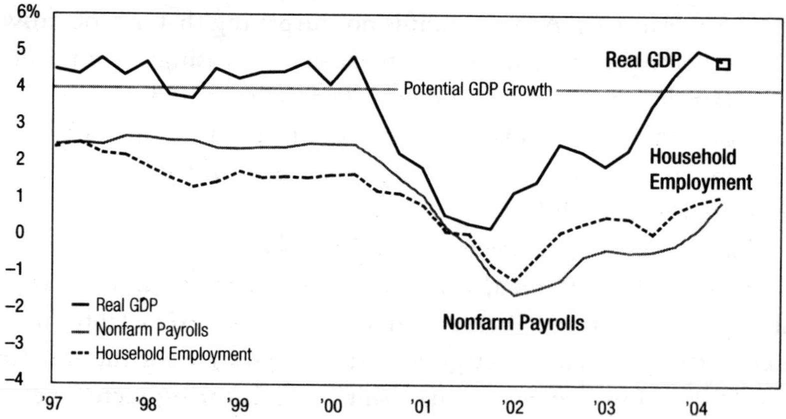

Source: U.S. Department of Commerce and U.S. Department of Labor

FIGURE 1.2 **Unprecedented Productivity Surge**
Annualized Percentage Change

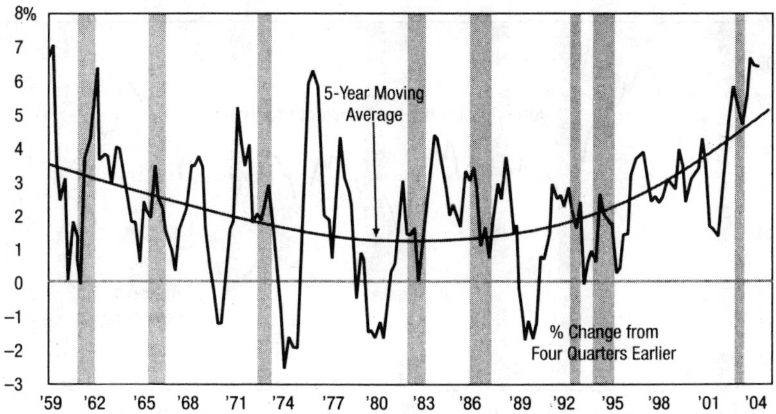

Source: U.S. Department of Commerce

far in excess of trend gains—before receding (see ***Figure 1.2***). The reasons for faster underlying productivity growth appear to reflect the direct and indirect benefits of a decade or more of outsized information technology investment. But cyclical factors have produced even stronger actual productivity gains as a result of efforts to improve efficiency and profits in an increasingly competitive global marketplace. Naturally, the successes of the corporate sector in reducing unit cost have come at the expense of cautious hiring.

The rapid productivity gains of the past few years have had a salutary impact on corporate profit growth. In fact, corporate profit margins have jumped to modern records (see ***Figure 1.3*** on the following page).

While there is no inherent reason why margins couldn't continue improving, the implications of further improvement would be particularly striking: The implied improvement in profits' share of GDP would move well beyond previous record highs, exceeding even optimistic expectations by a wide margin. Although the distribution of income varies somewhat over the course of the business cycle, income shares, including profit margins, tend to be relatively stable over broad spans of time. The profit share usually falls sharply during business slowdowns, when firms are slow to resize their operations, and recovers in the early years of expansion because companies are slow to recognize the improvement in the economic environment. With profits

FIGURE 1.3 **Profit Shares Move into Record Territory**
Ratio to Nominal GDP

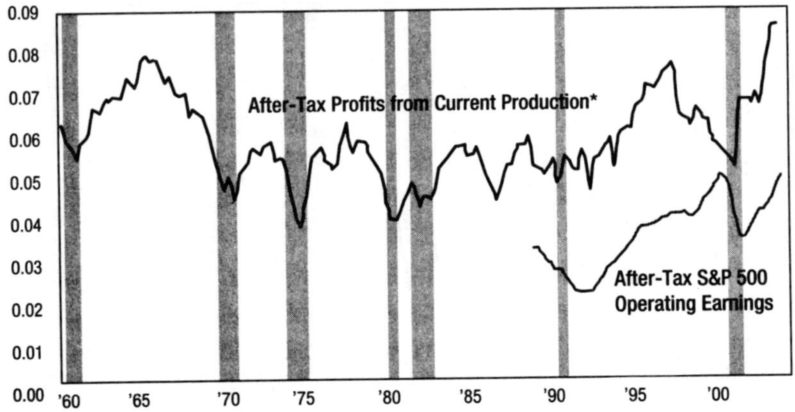

*With inventory valuation and capital consumption adjustments

Source: U.S. Department of Commerce and Standard & Poor's

already at a record high, it seems reasonable to expect that margin expansion ought to begin to stabilize.

The most likely course of events is that as business optimism builds and companies begin building capacity in anticipation of growing sales, profit margin expansion will stabilize. Part of the process of adding capacity will likely be an increased willingness to add workers. If this did not happen—that is, if productivity continued to advance at the exceptional pace of the past year or so, the implication is that profit margins would soar, despite a starting point of record highs. Such an outcome is not impossible, of course, but it seems improbable. In fact, the net job gains that appeared during the second quarter of 2004 signaled that the anticipated outcome is unfolding.

Increased Productivity Is the Key to Increased Compensation

Recent worries about the quality of newly created jobs is surprising. Naturally, the pickup in new hiring is encouraging because it indicates that businesses are becoming more confident in the prospects for sustained expansion. However, few workers stay in their first job. As a result, the initial job opportunity represents a first foot in the door that almost always leads to new opportunities. Thus, the combination of improving economic prospects and low inflation—together with strong productivity gains—point to an exceptionally promising outlook for workers.

In broad terms, faster productivity growth holds the key to improving job quality and increased compensation. Over time, productivity gains always accrue as increased labor compensation, reflecting a broadly competitive labor market. The recent surge of profit margins to modern records suggests that the recent productivity advances have not been fully anticipated. As the recovery advances, it is very likely that labor compensation trends will strengthen as well. Indeed, the distribution of national income between profits and labor has been relatively constant over long periods of time (see ***Figure 1.4***).

Recent disagreements about labor income gains reflect confusion about the various measures of labor compensation. Average hourly earnings of production and nonsupervisory workers, reported monthly from the survey of business establishments, indicates that wages have slowed to a 2 percent year-over-year rise—no faster than the rise in inflation over the past year. Of course, the unexpected rise in inflation early in 2004 reflected transitory factors, such as the surge in energy prices. Subsequent slowing in inflation will imply less of a toll on hourly wages.

FIGURE 1.4 **Workers Always Benefit from Faster Productivity**
Annualized Percentage Change Over Span Shown

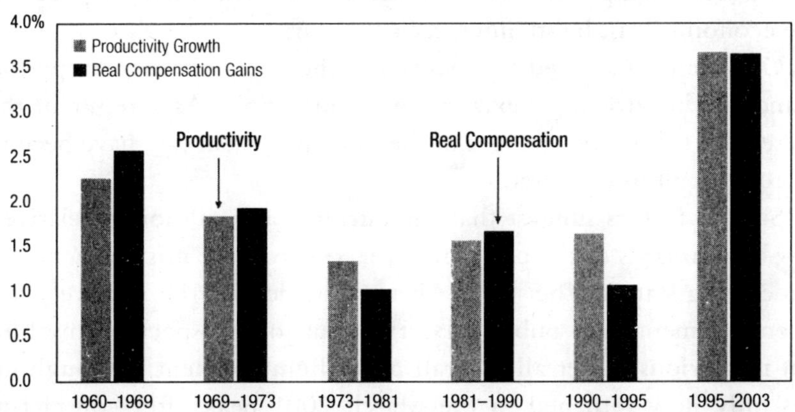

Source: U.S. Department of Commerce

However, monthly average hourly earnings figures represent only a portion of total labor compensation, because they exclude the compensation of managers and many professional occupations. In addition, average hourly earnings do not include commissions earned by sales workers. Finally, average hourly earnings exclude employee benefits that represent about 25 percent of total labor compensation.

The Employment Cost Index, which provides a broader perspective of worker compensation, indicates that since 2000 civilian compensation has been rising annually about 3.5 to 4 percent, significantly faster than inflation. This trend consists of benefit cost increases of about 7 percent annually plus 2.5 percent wage rises. If underlying productivity continues to grow by about 3 percent annually, labor compensation would be expected to grow by about three percentage points faster than inflation, as the recovery becomes more balanced. Against that backdrop, recent worries about the quality of new jobs are premature.

Returning Discouraged Workers Will Restrain Unemployment Decline

Visitors from the recent past would be confused by current economic events. For example, the lack of job growth—despite rapid GDP gains—would seem perplexing. At the same time, our time-traveler would wonder why there was so much public anxiety about the economic outlook, especially with the unemployment rate only at 5.5 percent (see ***Figure 1.5*** at right). At the very least, an unemployment rate this low would be presumed by our visitor to create a risk of accelerating inflation, as a 5.5 percent unemployment rate in past decades would have implied that the economy must be straining capacity limits.

One element needed to understand the economy's recent performance is the growing flexibility of labor supply. As a result of this increasing flexibility, changes in the unemployment rate have become more difficult to interpret.

Several factors suggest that the current rate—although relatively low by standards of the past thirty years—probably misrepresents the underlying state of the job market. First, despite the relatively low unemployment rate, public anxiety about job prospects is anything but mysterious. After all, overall payroll employment, although up, has only now returned to the March 2001 peak after contracting by two million in the slowdown/recession that began in mid-2000.

FIGURE 1.5 **Where Are We in the Life Cycle of the Economic Business Cycle**

Selected Unemployment Rates, Percent of the Labor Force

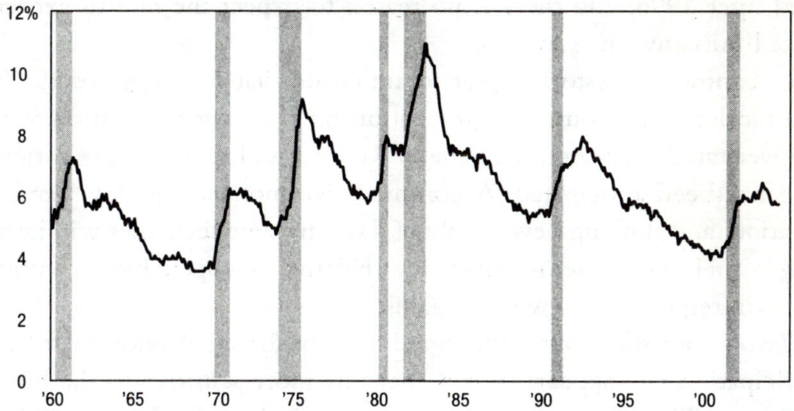

Source: U.S. Department of Labor, Bureau of Labor Statistics (CPS); Monthly Data

Second, the wide gap between output growth and new hiring is a reminder that the economy has demonstrated greater productive capacity than commonly assumed. Moreover, businesses continue to add to productive potential by investing record amounts in information technology and other capital equipment.[1] Third, benign inflation and falling unit labor costs—confirmed by record profit margins—suggest that the nation's productive resources remain underutilized. Finally, the unemployment rate never climbed as much as would have been expected, because growing numbers of young people—particularly those under 25 years old—probably are returning to school as they perceive limited immediate job opportunities. Therefore, the recent slack labor force growth reflects cyclical factors rather than secular demographic changes.

In other words, the limits to growth appear to be more flexible than commonly believed. Furthermore, there are no simple rules of thumb—such as a threshold unemployment rate—for judging maximum sustainable activity. Finally, despite the relatively low unemployment rate—traditionally the most reliable indicator of economic vitality—there is no convincing evidence that the U.S. economy is at or near binding capacity constraints.

Several implications of these considerations do not appear to be reflected fully in financial markets. If capacity limits remain distant, there are good reasons to expect inflation to remain benign for some time. In fact, the Federal Reserve's favored inflation measure—the core personal consumption expenditure chain-type price index measured on a year-over-year basis—has remained within a 1–2 percent band since 1996, and there is no reason to expect the rate to exceed those limits anytime soon.

In contrast, investors appear to anticipate that inflation eventually will accelerate to around 2.5 percent or so.[2] If investors' inflation fears are overstated at present, the Federal Reserve will remain more patient than had been anticipated. Accommodative monetary policy, benign inflation, and slowing new supply of fixed-income securities will keep long-term interest rates relatively low, helping to support higher equity prices in response to new profit gains.

Two points stand out with regard to labor force behavior. First, the participation rate appears to have become more sensitive to the business cycle. The effect appears to be symmetrical, in that both the "discouraged worker" effect of downturns and the "encouraged worker" impact of upturns have become more prominent than in the past.[3] Second, there has been a secular trend toward higher participation rates—if the age distribution of the population is held constant. In general, the rise in women's participation rate has outstripped the tendency for men's rates to decline, especially for older male workers.

Over the longer run, other factors also will influence the actual participation rate. First, the age distribution of the population will change. On the one hand, the aging of the baby boomers will tend to lower participation rates, although increases in life expectancy are likely to raise participation rates for older age cohorts. At the same time, immigration tends to lower the age distribution while raising participation rates. Thus, the apparent post–9/11 slowdown in immigration will have lowered both labor force growth and the participation rate; but this slowdown likely will prove to be temporary, as immigration has reaccelerated already to match the pre–9/11 pace.

In concrete terms, the recent decline in the labor force participation rate is greater than that experienced in the past (see ***Figure 1.6*** at right). This trend has been associated with considerably slower growth in the labor force than in employment, especially over the past year. As a result, the contraction in the number of people looking for work has resulted in a notable decline in the unemployment rate in the past year, even though employment has expanded only modestly.

FIGURE 1.6 **Dropouts Rise When Job Opportunities Are Limited**
Annualized Percentage Change

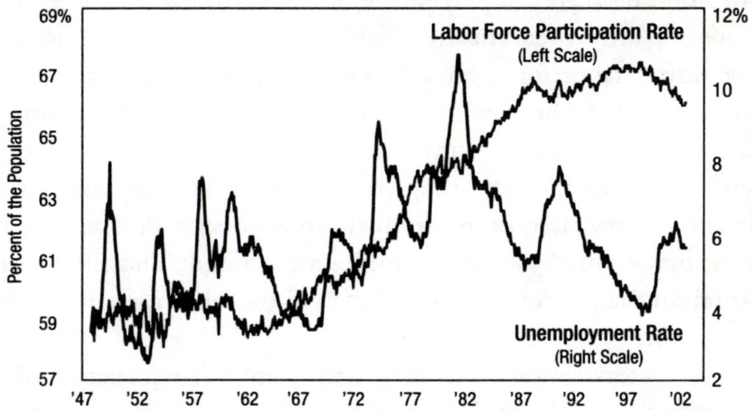

Source: U.S. Department of Labor

In this case, the new flexibility of the "supply" of new workers has diminished the usefulness of the unemployment rate as an indicator of pressure on labor resources. If labor supply is more endogenous—that is, has become more sensitive to the strength of the labor market—then declining unemployment may not be a sign that labor markets are tightening. In fact, a stronger job market would be expected to lure new workers, resulting in a stable rate, or possibly even a period of an increasing unemployment rate as the job market improves.

School-Age Dropouts Are a Cyclical (Temporary) Development

Understanding whether the recent drop in the participation rate is a result of mainly structural or cyclical forces is critical to interpreting recent developments and their implications for future inflation risks. Anticipating future shifts in the rate will influence judgments about the degree of slack in the economy. Thus, this issue will likely prove to have an important impact on future Fed policy, as well as on the outlook for long-term interest rates.

It seems reasonable to conclude that structural factors are not the principal explanation for the labor force growth or for the decline in the labor force participation rate in recent years. Immigration can be ruled out as a dominant factor. For sure, immigration slowed in the aftermath of the September 11, 2001, terrorist attacks. Nonetheless, the number of applications for immigration benefits has accelerated again, and the list of pending applications is mounting to a record number.

A decision by a greater percentage of adult women to stay at home and raise a family theoretically could provide a structural explanation for slower labor force growth. However, the latest data indicate that more men have been pulling out of the labor force than women. This trend contradicts the conventional wisdom that because young women have more choices—they can choose to work or raise a family, or both—they may be more likely than their male counterparts to stay at home when job opportunities are limited. Thus, the recent secular fall in the participation of women in the labor force is likely to be temporary.

Cyclical factors almost certainly are the principal explanation of the recent decline in the labor force participation rate. This interpretation is suggested by the divergent trends by age group. The decline in the participation rate of young adults aged 25 years old or younger in recent years is the most striking—and telling—signal that cyclical factors are the principal explanation for the slow growth of the labor force. After all, this age cohort is the principal source of new workers.[4] Strikingly, teenagers' participation rate has been falling at the fastest pace in the entire post–World War II period from 54 percent several years ago to 44 percent currently. Similar trends are evident for 20 to 24 year olds

FIGURE 1.7A **Participation of 20- to 24-Year-Olds**
Percent of the Population

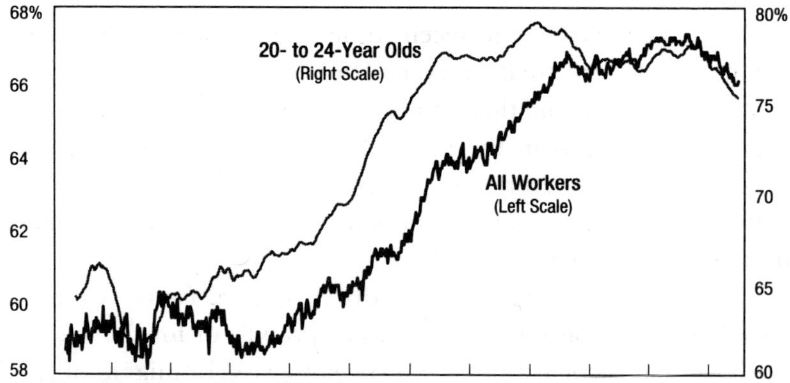

Source: U.S. Department of Labor

(see ***Figures 1.7a*** and ***1.7b***). These same developments were observed a decade ago. Subsequently, they were followed by a gradual return to the labor force as the economy rebounded.

Naturally, fewer women of the prime child-bearing age range of 25 to 34 years participate in the labor force than do men of the same age, but even in this group the differences between men and women have been converging sharply since the 1960s. Although demographers have speculated for some time that this convergence will soon stabilize, there is little evidence of this as yet. For example, the female participation rate for this age cohort stalled during the 1990 recession and has declined since 2000. Nonetheless, these two disruptions in the trend almost certainly were exacerbated by cyclical factors.

A widely noted development in recent years suggests that the secular rise in the female labor force participation rate since the 1960s is likely to continue. In fact, there is substantial upside for female participation rates, with the male-female gap still quite significant. At present, a greater number of women are going to college than are men. This is a strong indication that female participation in the labor force will continue to grow because those who are college-educated have a higher probability of participating in the labor force than those with less education: Specifically, 77.7 percent of college graduates participate in the labor force; 72.1 percent of those with some college experience participate; 63.3 percent of high school graduates with no college

FIGURE 1.7B **Men and Women ... Similar Trends**
Annualized Percentage Change

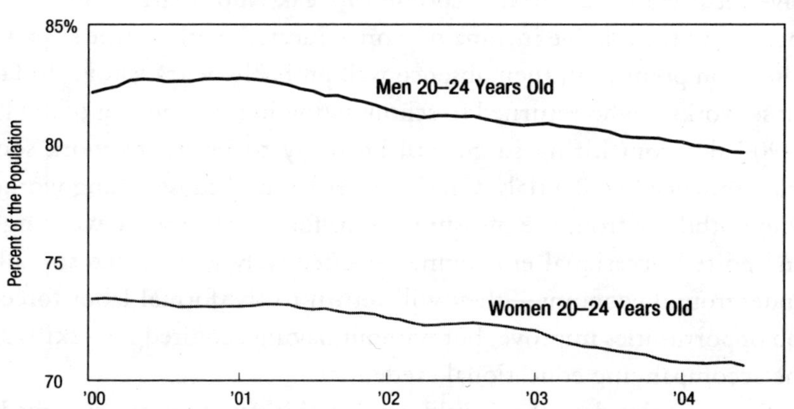

Source: U.S. Department of Labor

training participate; and 45.4 percent of those with less than a high school training participate in the labor force. In addition, the rise in voluntary part-time work—indicating that employers are providing workers with a wider mix of options—also has tended to help boost female participation rates.

In other words, the recent decline in the female labor force participation rate almost certainly reflects cyclical rather than structural, demographic, forces. This conclusion tends to be reinforced by the work experience of older workers as well. As a result, workers are likely to return to the job market as the economy strengthens and job opportunities broaden.

It has been surmised that the 1990s' rise in household net worth associated with rising asset prices would tend to encourage early retirement. However, the data do not indicate that this is taking place. Otherwise, some of the recent decline in the overall participation rate would probably not be reversed in coming years as the job market improves.

To the contrary, the evidence points to rising participation rates of older workers. For example, the participation rates of 45 to 54 year olds, men and women, are stable or climbing. At the same time, workers close to retirement, ages 55 to 64, appear be staying longer on the job. The same is true for even older workers. However, this effect is consistent with two developments: the increase in life expectancy and the decline in the relative importance of manual labor.

The decline in the labor-force participation rate for those no older than 25 years represents something of a puzzle. If young workers have returned to school in response to weak labor markets and to the increase in the relative income of more educated workers (the so-called education premium), their absence will probably be transitory. In fact, those workers who returned to school following the souring of the late 1990s' dot-com hiring surge will be ready to return to work soon, with enhanced credentials. On the other hand, if those young workers who withdrew from the measured labor force in fact were working in unreported occasional employment—effectively joining the so-called underground economy—they will return to the formal labor force as job opportunities improve, but without having acquired new skills and the accompanying educational credentials.

Surprisingly, data that could resolve this question are not readily available. Statistics on the number of students who are in college, graduate school, law schools, medical and dental schools, or business schools are sketchy or dated. Federal Reserve chairman Alan Greenspan noted in a recent speech that enrollments in community colleges have not increased noticeably. Other figures indicate the opposite: Figures from the Council of Graduate Schools show that enrollments grew 2 percent in 2000, following annual gains of 1 percent between 1986 and 2001. Moreover, some universities report 50 percent enrollment increases in continuing education and professional studies.[5]

In conclusion, the effective pool of available workers—including those who have gone back to school—is probably substantially larger than indicated by the government's current labor force statistics and its estimate of discouraged workers as well. If young people have gone back to school, they will not be included even in the government's estimates of discouraged workers. Therefore, as the job market improves and the work force expands in response, the economy's potential output will grow as well.

The job market is finally on the mend. Of course, the U.S. economy has a long way to go before it returns to full employment. However, evidence that hiring is advancing at an above-trend rate is muting earlier fears that the economy could not continue to recover on a sustained basis until a full-blown jobs recovery was under way.

CHAPTER NOTES

1. See "US Investment Surge and the Fed: Capacity Growth Lowers Inflation Threat," *Global Issues,* March 2, 2004, JPMorgan Securities Inc.

2. At mid-2004, the differential between ten-year conventional and inflation-protected (TIPS) Treasury yields had climbed to 2.75 percent. The TIPS asset class may give a distorted signal of inflation expectations because this market is relatively new and somewhat illiquid. Moreover, new Treasury offerings are lagging the growing universe of investors, who were restricted from investing in TIPS until they had a track record of at least five years of performance. Nonetheless, a number of long-standing surveys of investors' inflation expectations—the Livingston survey, the Federal Reserve Bank of Philadelphia survey of professional forecasters, the University of Michigan surveys, and the Blue Chip Consensus of private forecasters—indicate that ten-year CPI inflation expectations have not come down in this business cycle, even as core consumer inflation has broken down from around 3 percent to 1 percent.

3. The Labor Department's estimate of the pool of available workers provides a more expansive estimate of the potential supply of labor. It includes people who not only are unemployed and actively looking for a job but also have become discouraged and are not searching for a job actively enough to be considered in the labor force. In this sense, the

pool of available workers represents a narrow estimate of the potential supply of workers, because it excludes folks who couldn't find a job and decided instead to go back to school for more training.

4. See "Beyond that Jobless Rate: It's Down as the Young Stop Looking for Work," *BusinessWeek,* March 15, 2004.

5. Thanks to Courtney Schlisserman, Bloomberg News, for this information. Figures on the number of students taking admissions tests, including the GMATs, MCATs, GREs, and LSATs, may provide an indirect estimate of these trends, but these figures are not easily available.

2

The Global Labor Arbitrage

STEPHEN S. ROACH

STEPHEN ROACH insists that there is value in looking out beyond next year's economics. As chief economist at Morgan Stanley, Roach is, to be polite, concerned about economic outcomes that, to many of us, reside in the distant future. Roach brings a prodigious set of capabilities, including a political and economic tool kit that makes him the *pessimist (realist?) worth reading for optimists worldwide. Roach forces our attention. Here, he assesses the productivity and the information-technology-enabled efficiency of a future world economy. He delivers an overarching essay, impatient with our inability to confront, consider, and to finally come to terms with what lies ahead.*

The state of the U.S. labor market has been the defining issue of the current macroeconomic debate. During this period an unprecedented hiring shortfall has crimped the economy's income-generating capacity as never before. The American consumer, lacking in organic growth of purchasing power, has turned instead to riskier sources of support—namely, the combination of tax cuts and the debt-intensive extraction of home equity. The hope all along was that a standard cyclical recovery in job growth would finally kick in, thereby putting the United States on a more solid recovery path. While there has been some improvement on the hiring front over the past year, the quality of such job creation has been decidedly subpar. Unless that changes, the risks to a sustainable economic recovery in the United States will only intensify.

The extraordinary sluggishness of U.S. labor market activity is an outgrowth of one of the most distinctive features of the current economic climate: the cost-cutting tactics of the global labor arbitrage. In the absence of meaningful pricing leverage, cost cutting remains an absolute imperative for high-cost countries like the United States. With labor accounting for the bulk of Corporate America's cost structure, the global labor arbitrage has become central to strategies of competitive survival. At work are two of globalization's most powerful forces—vast offshore labor pools and the Internet. With the click of a mouse, the labor content of goods-producing and services-providing activities can now be extracted from remote locations at considerable saving. For high-cost economies like the United States, that puts extraordinary pressure on both head count and real wages. In many respects, the global labor arbitrage turns the world inside out.

Jobless Recovery—Quantity and Quality

Perspective is key to understanding the unique character of the current hiring cycle. For the first twenty-seven months since the inception of this recovery, America was mired in the depths of the worst jobless recovery of the post–World War II era. Then at long last, the magic seemed to be back as hiring picked up in the spring of 2004. In the twelve months ending May 2005, U.S. businesses added nearly 1.9 million workers to private nonfarm payrolls—an average of 156,000 per month. Such vigor was last seen in the year of the Great Bubble—specifically, back in September 2000, when the twelve-month gain in private nonfarm payrolls was running at a 2.2 million clip. While that increment stands in contrast to the net loss of 413,000 jobs in the first twenty-seven months of this recovery, it hardly breaks the mold of the weakest hiring cycle in modern history. Indeed, from the trough of the last recession in November 2001 through May 2005, private nonfarm payrolls rose a paltry 1.9 percent. This stands in sharp contrast to the 11.2 percent increase recorded, on average, over the same forty-two month interval of the five preceding recoveries (see ***Figure 2.1***).

While recent job gains have been impressive, they have not exactly been concentrated in the cream of the occupational hierarchy. Industries leading the pack on the hiring front over the past year include (in descending order): construction and real estate (311,000), health care and social assistance (298,000), administrative and waste services

FIGURE 2.1 **America's Jobless Recovery**

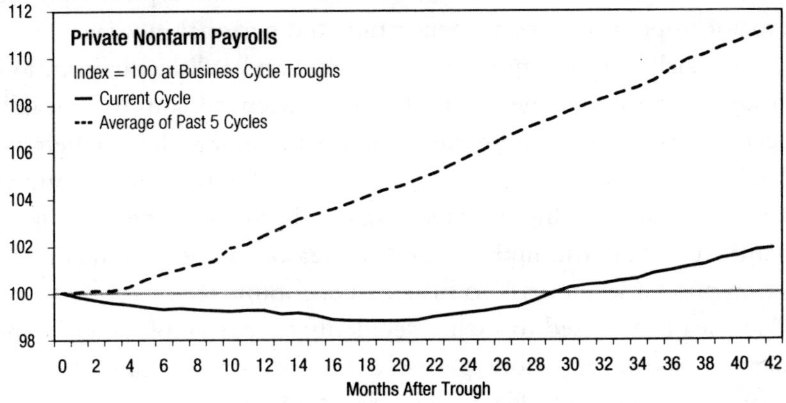

Source: Bureau of Labor Statistics and Morgan Stanley Research

(234,000), and restaurants (229,000). Collectively, these four industry groupings, which employed 36 percent of all U.S. workers on private nonfarm payrolls a year ago, accounted for nearly 60 percent of the total growth in private hiring over the most recent twelve-month period. Apart from the obvious impact of the housing bubble on relatively high-wage employment in real-estate-related activity, the industry mix of the hiring dynamic remains skewed toward the lower end of the U.S. pay structure.

That's not to say there hasn't been any improvement at the upper end of the U.S. labor market. There have been signs of improvement in several of the higher-end professional services categories—namely, accounting, architecture and engineering, computer systems design, consulting, credit intermediation, the brokerage and securities industry, and private education. Collectively, this latter group of industries, which makes up another 10 percent of overall employment, accounted for 18 percent of total job growth over the past twelve months. Overall, as seen from the standpoint of this industry-by-industry breakdown of the U.S. job structure, there can be no mistaking the bifurcation of the improved hiring dynamic during the most recent twelve-month period. The contribution of lower-end jobs (60 percent) was more than three times that of higher-end jobs (18 percent)—which qualifies as a decidedly low-quality improvement in the U.S. labor market.

Admittedly, there has been a pickup in the pace of U.S. job creation this past year, but the bulk of the impetus—albeit an unusually anemic one by cyclical standards of the past—has been concentrated at the low end of the quality spectrum. The Great American job machine is not even close to generating the high-powered jobs that typically provide the major impetus to income generation and personal income.

This conclusion has important economic and political implications. In response to the income shortfall, overly extended consumers can be expected to go further out on the risk curve in order to defend their lifestyles. The Fed, for its part, will be more wary of normalizing monetary policy if it means that higher interest rates will threaten the asset-driven dynamic to U.S. consumption. For those reasons alone, low-quality job creation poses a serious risk to sustained economic recovery.

America is not used to such a decidedly subpar employment experience: it has never before persisted for forty-two months into an economic recovery. A decade ago, the United States went through its first so-called jobless recovery. After a painful and unusual wait of nineteen months, the hiring cycle turned sharply to the upside in late 1992 and then never looked back. The current experience is far more extreme on a variety of counts—overall hiring, the quality of jobs, and the ongoing compression of real wages.

Wage and Income Compression

That brings up a second and equally worrisome dimension of America's labor market conundrum—a decided shortfall of real wage growth. This is critically important for the consumer because personal income generation is driven by the interplay between employment and the pay rate. Monthly data on average hourly earnings in March and April 2005 hinted at a long-awaited revival in wages—average gains of about 0.3 percent at a monthly rate, or close to a 3.4 percent annual rate. The May number—a fractional increase of just 0.2 percent—draws that optimism into serious question. This series is now up just 2.6 percent on a year-over-year basis through May 2005, well below the 3.5 percent increase in the headline Consumer Price Index (CPI) in the twelve months ending in April and slightly above the 2.2 percent rise in the core CPI over the same period.

The lack of improvement in real wages, in conjunction with the hiring shortfall, has put a real squeeze on the internal generation of consumer purchasing power. The wage and salary component of personal income is currently $258 billion below the average profile of the

FIGURE 2.2 An Income-Short U.S. Economy

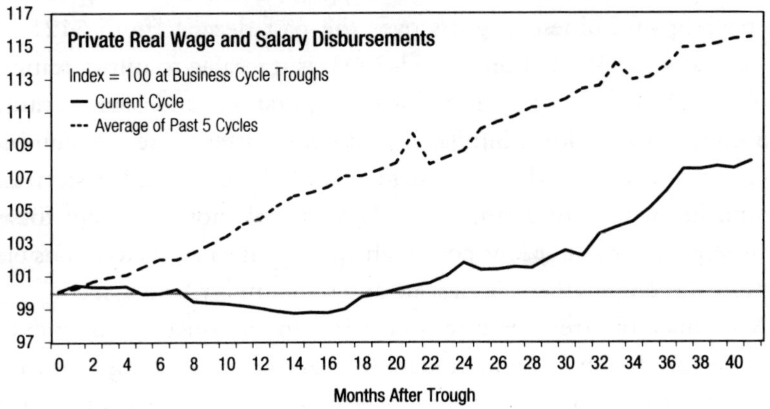

Source: Bureau of Economic Analysis and Morgan Stanley Research

past five recoveries (see *Figure 2.2*). (Note: This comparison only goes through April 2005.) Little wonder then that income-short American consumers have morphed into asset-dependent spenders.

Behind the Arbitrage

We hear repeatedly that the disconnect in the U.S. labor market is all about lags or productivity. I don't buy it. Instead, I believe that a new force has come into play that is now altering the fundamental relationship between domestic demand and domestic employment in the United States. I call it the *global labor arbitrage*—the IT-enabled efficiency tactics that allow U.S. companies to replace high-wage domestic workers with like-quality low-wage foreign workers in goods-producing and services-providing functions alike. The lack of pricing leverage in today's climate makes this arbitrage an increasingly urgent competitive imperative. The global labor arbitrage is likely to be an enduring feature of the macro climate, raising the distinct possibility that subpar job creation and muted gains in real wages could well be here to stay in the United States for the foreseeable future.

A unique and powerful confluence of three mega-trends is driving the global labor arbitrage: the maturation of offshore outsourcing platforms, e-based connectivity, and the imperatives of cost control.

1. The maturation of offshore outsourcing platforms—China exemplifies the critical mass in new manufacturing outsourcing platforms. Built on a foundation of massive inflows of foreign direct investment and domestically funded infrastructure, the Chinese factory sector has become a key link in the global supply chain. Sixty-five percent of the tripling of Chinese exports over the past decade (from $121 billion in 1994 to $365 billion in mid-2003) is traceable to outsourcing by Chinese subsidiaries of multinational corporations and joint ventures. And China is not alone. Similar outsourcing patterns are evident elsewhere in Asia and in Mexico, Canada, South America, and Eastern and Central Europe. Outsourcing is hardly a new phenomenon, but today's offshore platforms offer low-cost, high-quality alternatives to goods production and employment on a scale and scope never before seen.

A comparable trend is now emerging in the once sacrosanct services sector. Dubbed "non-tradables," services have long been perceived as having to be delivered in person, on site. That's no longer the case. Offshore outsourcing of services is now occurring up and down the value chain—from low-value-added transactions processing and call centers to activities with a high intellectual capital content, such as software programming, engineering, design, accounting, actuarial expertise, legal and medical advice, and a broad array of business consulting functions. India exemplifies the critical mass in offshore services outsourcing. One study estimates that India's IT-enabled services exports will increase ten-fold over the next four years, from $1.5 billion in 2001–2002 to $17 billion by 2008, making it one of the fastest-growing major industries in the world (see *The IT Industry in India: Strategic Review 2002,* published by India's National Association of Software & Service Companies with McKinsey & Co.). India is not alone: Services outsourcing is increasingly prevalent in countries such as China, Ireland, and Australia.

2. E-based connectivity—This is the first business cycle since the advent of the Internet. Say what you will about the Web, but I believe it has transformed the supply side of the global macro equation. For manufacturing, it gives new meaning to the real-time monitoring of sales, inventory, production, and delivery trends that drive the logistics of global supply chain management. And it provides new transparency to the price discovery of factor inputs and upstream materials and supplies. For services, the Internet enables a dramatic expansion of outsourcing options. The intellectual capital of research, analysis, and consulting can be transmitted anywhere with the click of a mouse. A systems problem in New York, for example, can now be quickly fixed by a software patch written in Bangalore. Such connectivity creates a new pipeline for global information flows that drive the service-sector supply chain. The Internet allows well-educated, hard-working, and relatively low-wage offshore knowledge workers to be seamlessly integrated into global service businesses, once the exclusive domain of knowledge workers in the developed world.

3. The new imperatives of cost control—This trend is, in effect, the catalyst that brings the global labor arbitrage to life. In an era of excess supply, companies lack pricing leverage as never before. Therefore, businesses must be unrelenting in their search for new efficiencies. Not surprisingly, the primary focus of such efforts is labor, representing the bulk of production costs in the developed world. In the United States, for example, worker compensation still makes up more than 75 percent of total domestic corporate income. And that's the point: Wage rates in China and India range from 10 percent to 25 percent of those for comparable-quality workers in the United States and the rest of the developed world. Consequently, offshore outsourcing that extracts product from relatively low-wage workers in the developing world has become an increasingly urgent survival tactic for companies in the developed economies. Mature outsourcing platforms, in conjunction with the Internet, give new meaning to such tactics. General Electric's "70-70-70" credo says it all: One of the world's most successful companies has publicly stated the goals of outsourcing 70 percent of its headcount, pushing 70 percent of that outsourcing offshore, and locating 70 percent of such workers in India. With 16,000 workers in India today—about 5 percent of its global workforce of 313,000—GE has only just begun to exploit global labor arbitrage to achieve efficiencies in today's intensely competitive climate. This suggests that such an arbitrage is only in its infancy.

These mega-forces are largely irreversible, especially mature outsourcing platforms and the Internet. The imperatives of cost cutting could diminish once global supply and demand are in balance; but that is not likely to occur for some time. Meanwhile, the resulting global labor arbitrage continues to have a profound impact on job creation in the United States. Through mid-2005, private nonfarm payrolls remained about 10 million workers below the hiring trajectory of a typical economic recovery.

Halfway around the world, there are clear indications of complementary adjustments in Asia's huge reservoir of surplus labor. In China, foreign-funded subsidiaries now employ some 3.5 million workers, up more than 3.5 times over the past decade. Moreover, another 3.25 million Chinese workers are employed by subsidiaries funded in Hong Kong, Taiwan, and Macao. Similar trends are evident in services outsourcing. India currently employs about 650,000 professionals in IT services, a figure that is expected to more than triple over the next five years, according to the McKinsey study cited above. Moreover, there is good reason to believe that increased staffing by Indian subsidiaries of multinational service providers will be matched by headcount reductions elsewhere in their global platforms.

Two Different Models

As the global economy tilts increasingly toward Asia, the rest of the world struggles to cope. China and India are leading the way in driving the new global growth dynamic, but their approaches are very different. In China, manufacturing clearly has led the way. This transformation has been especially dramatic in recent years. The industrial sector's share of Chinese GDP rose from 41.6 percent in 1990 to 52.9 percent in 2004. Putting it another way, such industrialization accounted for fully 55 percent of the cumulative increase in China's GDP over this fourteen-year period. India's development model is cut from a very different cloth. The industry share of Indian GDP has been essentially stagnant in recent years—holding at 27 percent of GDP over the 1990 to 2004 interval. As a result, industrial activity accounted for only 27 percent of the cumulative increase in India's GDP over the past fourteen years—literally half the contribution evident in China (see ***Figure 2.3***).

For services, it has been the mirror image. In India, the services portion of GDP increased from 41.1 percent in 1990 to 51.7 percent in 2004. Over this fourteen-year period, services accounted for 54 percent of the cumulative increase in Indian GDP growth. By contrast, the services share of Chinese GDP rose from 31.3 percent in 1990 to only 31.9 percent in 2004. Not only is China's services sector a much smaller slice of that nation's economy than is the case in India, but the growth dynamic of Chinese services has been especially weak. Over the most recent fourteen-year period, the expansion of China's services economy accounted for just 32 percent of the cumulative increase in overall GDP—just over half this sector's growth contribution in India.

China's approach is a classic textbook example of manufacturing-led development—yet the Chinese also have taken this model to a new

FIGURE 2.3 **The New Asia**

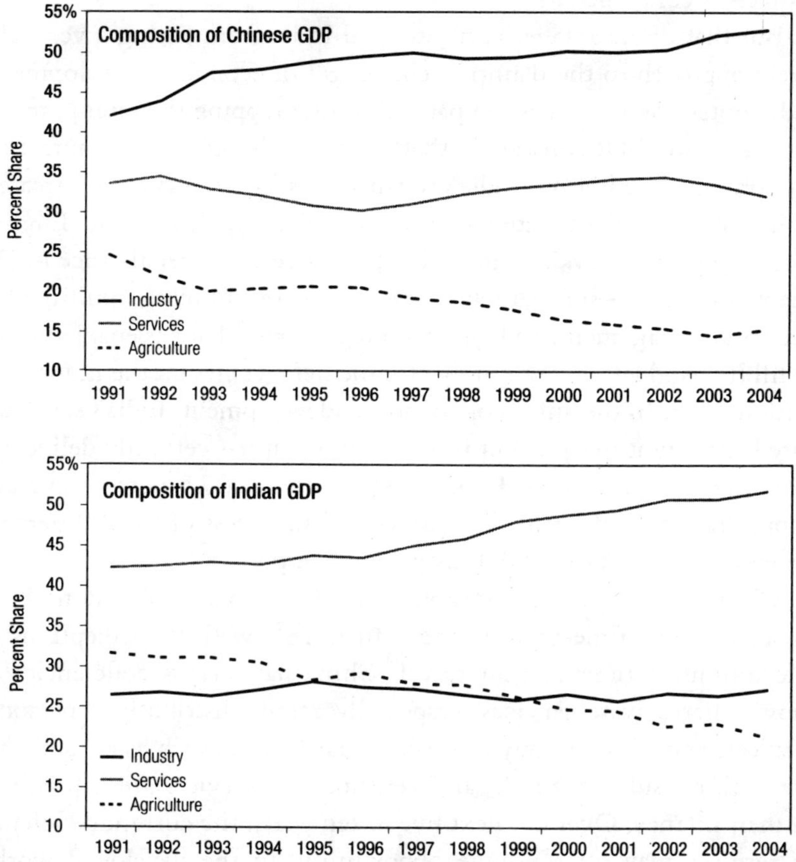

Sources: China National Bureau of Statistics, RBI, CSO, and Morgan Stanley Research

level. Four major factors appear to differentiate China's strain of industrialization from others—a 45 percent domestic saving rate, impressive progress on the infrastructure front, surging foreign direct investment (FDI), and a vast reservoir of hard-working, low-cost labor. While this progress has been remarkable for more than two decades, only in the past few years has China truly come of age as the world's factory. There also does not appear to be any let-up in sight. FDI of $60 billion surged into China in 2004—making this nation the largest recipient of such flows anywhere in the world. By contrast, India is at a major disadvantage on all counts: Its national saving rate of 28 percent is only a little more than half that of China's; its infrastructure is in terrible shape; and its ability to attract FDI—which ran at only $4 billion in

2003—pales in comparison with that of China. China may have at least a ten- to fifteen-year lead over India insofar as manufacturing prowess is concerned.

But that disadvantage hasn't stopped India from taking a very different approach to the daunting challenges of economic development. India opted for a services-led path, thus sidestepping the saving, infrastructure, and FDI constraints that have long hobbled its manufacturing strategy. In addition, India's reliance on services plays to its greatest strengths—a well-educated workforce, IT competency, and English language proficiency. The result has been a veritable renaissance in IT-enabled services—software, business process outsourcing, multimedia, network management, and systems integration—that has enabled India to fill the void left by seemingly chronic deficiencies on the industrialization front. In the annals of economic development, India's services-based strategy is unique, but in recent years, it has certainly delivered: The services segment of Indian GDP grew at a 7.5 percent average annual rate over the past five years, well in excess of the 5.9 percent average growth in total GDP over the same period.

China, for its part, is a serious laggard in services. You don't have to spend much time there to see it firsthand. With the exceptions of telecommunications and air travel, China has serious deficiencies in most other private services—especially retail, distribution, personal services, and a broad array of professional services such as accounting, medical, consulting, and legal. Even financial services are still largely in their infancy. Over the next five to ten years, the current deficiency in services may offer a huge opportunity. In the developed world, services account for at least 65 percent of total economic activity—double China's current share. Moreover, as reforms of state-owned enterprises continue to result in the elimination of 7–9 million jobs per year, the expansion of a labor-intensive services sector could fill an important employment need in China. For those reasons alone, there is nothing but upside to the Chinese services sector.

Services-driven development models, such as the one now at work in India, cast globalization in a very different light. Most important, they broaden the competitive playing field, thereby bringing new pressures to bear both on job creation and on real wages in the developed world. This is where the debate gets prickly. Protectionists scream, "Foul!"—arguing that trade barriers are the appropriate answer. Yet, in my view, there is nothing intrinsically unfair about these developments. Globalization is very much a moving target. The rules of globalization are dynamic, not static. They change as the world changes. The Asian challenge we now face may not be China *or* India—it may well be China *and* India.

The Great Offshoring Debate

The changing rules of globalization take the debate into its most contentious realm—the great dispute over offshoring. Most of the evidence on the effects of offshoring is circumstantial. For example, the 11.4 percent surge in U.S. real goods imports growth over the first six quarters of this recovery was far in excess of what might normally be expected in the context of an anemic 4.2 percent increase in domestic demand over the same period. In the case of the United States, rising import propensities and the concomitant offshore outsourcing of jobs are the functional equivalent of "imported productivity," as global labor arbitrage substitutes foreign labor content for domestic labor input. This could well go a long way in explaining the latest chapter of America's fabled productivity saga.

A review of the hard evidence on the impacts of offshoring doesn't take long. In large part that's because the actual data points on the empirical magnitude of offshoring are few and far between. Not surprisingly, the consultants—most of whom are in the IT advisory business—have tended to downplay the loss of jobs from the high-wage developed world to the low-wage developing world. The most widely cited estimate of the impact of offshoring comes from a study of U.S. trends conducted by Forrester Research (see the April 2003 Forrester study by Christine Ferrusi Ross, "Can Outsourcers Really Transform IT?"). It calculates that only about 400,000 business process jobs have been "offshored"—a total the firm expects to rise to about 3.3 million by 2015. That may sound like a lot, but it works out to annual job losses of only about 300,000 over the next decade—which is not much of a dent in a U.S. economy that currently employs 130 million workers.

As best I can tell, this is a pretty flaky estimate. Forrester does not provide much detail on the methodology or the empirics that lie behind this number. Moreover, it is important to keep in mind that the Forrester estimate pertains only to business process jobs—a relatively small slice of white-collar jobs that could ultimately be affected by IT-enabled offshoring. Unfortunately, a similar approach is taken by other IT consultants, even by those who think the macro impacts are a big deal (see Gartner's July 2003 research note by D. Morello, "U.S. Offshore Outsourcing: Structural Changes, Big Impact").

The best work I've seen on the so-called offshoring phenomenon has been produced by Catherine Mann of the Washington, D.C.–based Institute for International Economics (see her December 2003 IIE Policy Brief, "Globalization of IT Services and White Collar Jobs: The Next Wave of Productivity Growth"). Mann's approach is solid—relying on both analytics and empirics to develop a framework for assessing the impact of this phenomenon. But even she concedes that "there are no publicly available data on jobs 'lost' to workers in foreign economies." Nevertheless, she concludes by extolling the virtues of offshoring as yet another IT-enabled development that lowers operating costs of U.S. businesses. As a result, she maintains that such efficiency enhancements have the potential to provide U.S. companies with the wherewithal to expand hiring in the future. While it's hard to rule out such a possibility, there are many other conceivable outcomes. The bottom line, as I see it: We're largely flying blind in assessing the current and prospective magnitude of this important transformation in the U.S. labor market. My gut instinct tells me that this trend—like most IT-enabled developments in the past decade—is likely to proceed at a much faster pace than the consultants believe.

Politicization

Like it or not, this is the way globalization is supposed to work—which takes us to the toughest aspect of the problem: the distinct possibility that there may be strong social and political objections to the very concept of globalization itself. The idea that job contracts must be rewritten because of trade liberalization and an increasingly integrated borderless world doesn't sit terribly well with those disenfranchised workers who are on the front line of making the adjustment. Globalization may work well in the long run, but it appears to have profoundly disruptive impacts in the short run. That could reflect its inherent asymmetries—developing countries first come on line as producers long before they emerge as consumers. That leaves a very tenuous interregnum, where the creation of new markets in the developing world lags the penetration of old markets in the developed world.

And that's what takes us to the most dangerous point of all—the politicization of the great offshoring debate. In 2004, an election year, the American body politic was forced to take sides on this highly charged emotional issue. Analytics and empirics ring hollow in this deeply personal context. Free trade and now offshoring lie at one end of the spectrum, protectionism at the other. For America—complete with its jobless recovery and gaping trade deficit—the pendulum is now swinging in an ominous direction. China-bashing is on the ascendancy in Washington in the summer of 2005. In Asia, this is a huge and puzzling concern.

Disenfranchised workers take no comfort in the theoretical promises of globalization. The theory, of course, is that surging incomes in the developing world which arise from such offshoring spawn new markets and a new class of consumers. As supply begets these new sources of global demand, goes the argument, displaced workers in the developed world are presumed to be well positioned to uncover new sources of job creation. It's a fine theory, and over the long run inarguable, in my view; but the long run may be a good deal further in the future than most are willing to admit. First of all, consumers in low-wage developing nations such as China and India do not have job security or the benefit of institutionalized safety nets. China, for example, continues to eliminate 7–9 million positions a year under the guise of state-owned enterprise reforms and is lacking a national social security and pension system; little wonder then that its consumers remain predisposed toward saving. That underscores another one of the inherent asymmetries of globalization: the shifting mix of global job growth may initially be driven more by the supply side of the equation in the low-wage developing world; conversely, demand-side impacts, which might spur hiring in the high-wage developed world, could lag for a considerable period.

In the end, the global labor arbitrage may well meet its biggest challenge in the political arena. A record hiring and real wage shortfall in an election year certainly raises this issue to the top of America's political agenda; that's especially the case if job-related angst continues to move up the white-collar occupational hierarchy to middle-aged, high-skilled, upper-income segments of the U.S. workforce. These workers have long harbored the presumption of lifetime employment and have never before felt the pain of economic hardship and distress. Now they fear that their jobs are gone forever.

Harsh verdicts also are likely to be rendered by other politicians and policymakers around the world. Recent G7 communiqués are worrisome in that regard. Europe and Japan are now united in pointing the finger at China as the scapegoat of global rebalancing. They seem to believe that China must now bear a greater share of the impacts of a weaker dollar, a point of view that doesn't exactly sit well in Beijing these days. The global labor arbitrage has important consequences for geopolitical tensions as well.

Backlash Against Globalization

The IT-enabled global labor arbitrage is emblematic of the inherent contradictions of globalization: It is the means by which jobs are created in poor countries, while it is also the breeding ground of a political backlash in rich countries. Ultimately, these tensions will have to be vented—either through economics, or politics, or both. The steady drumbeat of America's jobless recovery tips the scales more toward the political resolution. For that reason alone, I continue to fear a backlash against globalization that takes the form of heightened trade frictions and mounting protectionist risks.

At the same time, the impacts of globalization are likely to be an increasingly big deal in driving the great American job machine in the future. This conclusion is best understood within the context of the mix of forces that drive turnover in the U.S. labor market—namely, the interplay between the constant flux of hirings and firings. The sum of these flows is, of course, considerably larger than the net changes that receive such great attention when the state of nonfarm payrolls is reported each month by the U.S. Bureau of Labor Statistics (BLS). For example, the BLS Business Employment Dynamics tabulation puts the sum of gross job gains and losses at some 15.4 million workers in the third quarter of 2004 (latest data point)—dwarfing the net change of 191,000 the BLS estimated for the quarter. According to Alan Greenspan, layoffs have not been the dominant force shaping America's jobless recovery; instead, it's the lack of hiring. He argues that "Gross separations from employment, two-fifths of which have been involuntary, are about what would be expected from past cyclical experience, given the current pace of output growth. New hires and recalls from layoffs, however, are far below what historical experience indicates" (see testimony of Chairman Alan Greenspan on the *Federal*

Reserve Board's Semiannual Monetary Policy Report to the Congress before the Committee on Financial Services, U.S. House of Representatives, February 11, 2004).

What matters most in shaping macro trends is change at the margin. The global labor arbitrage could well be having a differential effect on the gross flows in the U.S. labor market. It's not that domestic jobs are being eliminated on a large scale in the United States and shifted offshore to the developing world. Instead, it's far more likely that the impacts are being felt more on the hiring side of the equation. U.S. companies are now letting the "opportunity cost" of the domestic hiring decision be shaped increasingly by the alternative of highly educated, well-skilled, low-cost workers now readily available in many developing countries. At the same time, by lowering the perceived incremental cost of the "next hire" in many occupational categories, the arbitrage also could be playing an increasingly important role in the wage determination process that affects a much broader cross-section of American workers. In other words, the globalization of labor input doesn't have to be large in the absolute sense to make a difference in shaping change at the margin. The increased prevalence of offshoring suggests that such a critical mass may well have been attained in the United States. As a result, U.S. companies have no choice other than to become more global in both their perspective and structure.

Like most economists, I, too, worship at the high altar of free-market competition and the trade liberalization that drives it—but that doesn't mean necessarily putting a positive spin on the painful dislocations that trade competition can engender. That was the unfortunate mistake made in 2004 by the Bush administration's former chief economist, Gregory Mankiw, in his dismissive assessment of white-collar job losses due to offshoring. Like most economic theories, the optimal outcomes cited by Mankiw pertain to that ever-elusive long run. Over that time frame, the basic conclusion of the theory of free trade is inarguable: International competition lowers costs and prices, thereby boosting the purchasing power and standard of living of consumers around the world. The practical problem in this case—as it is with most theories—is the concept of the long run. Yes, over a long enough time frame things will eventually work out according to this theoretical script. But the key word here is "eventually"—the stumbling block in presuming that academic theories map neatly into the shorter time horizons of financial markets and politics. Lord John Maynard Keynes put it best in his 1923 *Tract on Monetary Reform,* cautioning, "In the long run, we're all dead."

For theorists like Mankiw, offshoring is seen as but a bump in the road. The presumption in this case is that an innovation-led, flexible U.S. economy is able to uncover new sources of job creation that can fill the void left by this cross-border labor arbitrage. Yet that may be a heroic assumption for the foreseeable future. As nontradables become tradable, America's once shielded white-collar workers face increasingly intense competition from increasingly well-educated foreign workers. And as skill sets converge around the world, the quick and seamless regeneration of hiring that underpins the theory of free trade starts to seem like an increasingly unrealistic assumption. It's not the theory of free trade that has been invalidated, as some have argued, such as New York Senator Charles Schumer (see Senator Charles Schumer and Paul Craig Roberts, "Second Thoughts on Free Trade," the *New York Times,* January 6, 2004). Ironically, it's that this theory now applies far more broadly than ever imagined.

American politicians certainly sense this undercurrent of angst in the U.S. labor market. The pro-labor mood in the Congress is both extreme and bipartisan. As one Capitol Hill veteran put it to me in the fall of 2003 when I was testifying on U.S.-China relations, "The protectionist train has left the station." While campaigning in Pennsylvania in the fall of 2004 President Bush said, "There are people looking for work because jobs have gone overseas. We need to act." His opponent, Senator John Kerry, expressed similar views. The real risk, of course, is that the politicians do the wrong thing. Bills already have been introduced in both chambers of Congress that would put steep tariffs on all Chinese products sold in the United States. There is talk of going after India and imposing tax penalties on U.S. multinationals that shift jobs overseas. Several states have introduced legislation that bans offshoring contracts. And U.S. immigration authorities have sharply reduced the cap on so-called H-1B visas that cover the entry of foreign IT workers.

The offshoring debate is not about to go away. Neither theory nor fact will temper the palpable sense of angst that has arisen in America's unprecedented jobless recovery. The drumbeat of protectionism grows louder at precisely the moment when the United States has a record current-account deficit, a weakening dollar, and an extraordi-

FIGURE 2.4 **Global Imbalances and Currency Adjustments**

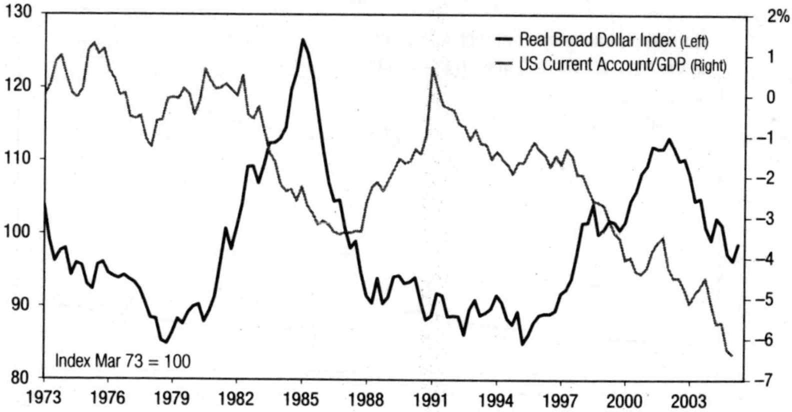

Source: Federal Reserve Board and Bureau of Economic Analysis

nary dependence on Chinese financing (see ***Figure 2.4***). It's a house of cards that has never seemed more precarious. Yet ever-complacent financial markets could care less. The risk is that they will—sooner rather than later.

What About Us?

Hiring cycles will always come and go. But as we can full well see in the experiences of Europe and Japan, new structural forces can come into play that have a lasting and profound impact on job creation. Globalization remains the most powerful economic force of the modern era. It was only a matter of time before the IT-enabled globalization of work had a major impact on the U.S. labor market—and that time is now. The character and quality of American job creation is changing before our very eyes—which poses the most important question of all: What are we going to do about it?

Alas, globalization and the economic development it fosters are a two-way street. If China continues to deliver in manufacturing and India pulls off a rare services-led development strategy, the wealthy industrial world will face new and important challenges. The theory of trade liberalization and globalization maintains that there is little to worry about. After all, in the long run, the income workers make as producers should show up on the other side of the ledger as purchasing power for consumers. As the developing world's fledgling consumers

FIGURE 2.5 **A Worrisome Human Capital Gap**

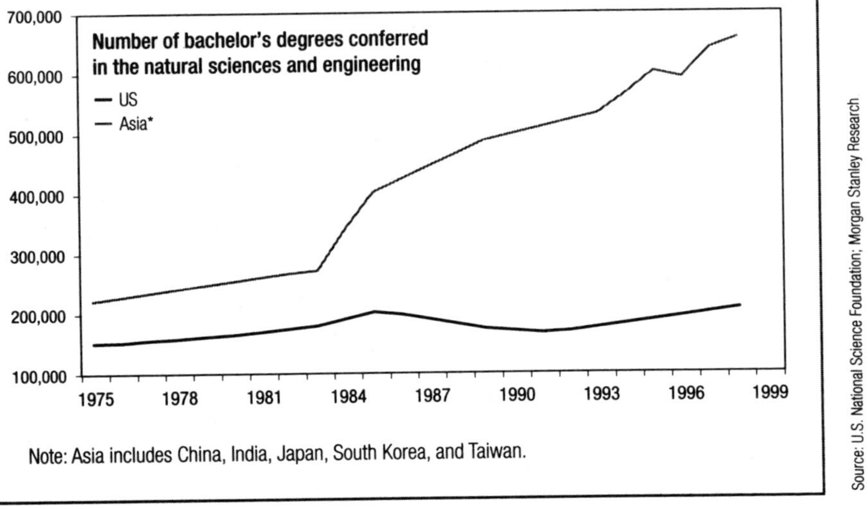

Note: Asia includes China, India, Japan, South Korea, and Taiwan.

then come to life, goes the argument, new opportunities and markets will be given to suppliers in the developed world. All this is potentially a big plus for the world economy. Globalization need not be seen as a "zero-sum" outcome.

From the standpoint of wealthy workers in the developed world, that raises another serious problem: a narrowing of the educational attainment gap between the developed and developing worlds. This could well inhibit the knowledge-based job creation that high-wage Western economies are counting on to fill the void of the cross-border labor arbitrage—a possibility that should not be taken lightly. U.S. National Science Foundation data show that the United States is currently awarding only about 200,000 bachelor's degrees in engineering and science, little changed from trends in the mid-1980s. By contrast, Asia's annual graduates of science and engineering students (China, India, Japan, South Korea, and Taiwan, combined) has now hit approximately 650,000 per year; that's up over 50 percent from the graduation rate in the mid-1980s and fully three times the comparable degree production rate in the United States (see *Figure 2.5*). The United States has long drawn comfort from the quality differential of its educational system. However, in the Internet Age with its ubiquitous diffusion of knowledge, innovation, and technological change, that may turn out to be an increasingly false sense of security. Needless to say, convergence on the human capital front raises serious questions about America's future competitive prowess, as well as its ability to uncover new sources of job creation.

As education and skill levels are raised around the world, and as the world itself is brought closer together through IT-enabled connectivity, the wealthy developed world must rise to the occasion. That means doing what we have always done best—staying open and flexible, and pushing the envelope on education, technological advancement, and risk-taking entrepreneurial activity. No one said it was supposed to be easy—but it sure beats the alternatives. The global labor arbitrage is a clear and important reminder of how tough the heavy lifting is likely to be.

3

America's Optimistic Future

DAVID R. MALPASS

DAVID MALPASS is optimistic. The Bear Stearns chief economist combines economics and his Reagan and Bush, Sr. administration experience with an original command of the language. The result, through times of worry and angst, is a relentless belief in the world economy's ability to heal, gather strength, and prosper. Unlike less capable "life is good" acolytes of the modern day, Malpass grounds his optimism in market-based analysis and backs it up with paragraph-chart-paragraph presentation of real economic data. Here, David Malpass nudges fear aside with a forecast of future mornings across the economic world. We will land on two wheels.

It is sometimes hard to see through the gloom of 9/11, the Iraq war, a divisive presidential election, and constant confusion about the economic outlook. Add to that the ups and downs of financial markets.

But I see reasons for optimism. Despite all the critics in 2002, the U.S. economy wasn't fragile. The recovery turned into an expansion and then a strong expansion. By 2004, year-over-year growth exceeded the peak in the late 1990s, a considerable achievement given the dire predictions of 2003.

Though it will take some years to test, I think the second half of this decade will show:

- A satisfactory end to the traumatic inflation/disinflation/deflation cycle from 1971–2001. The shift back to price stability should be good for economic growth, even if prices pass briefly through a period of moderate inflation. I expect a multiyear process of inventory rebuilding and above-trend increases in demand for commodities, with corresponding implications for profits. The yield curves should shift up toward more normal levels in the United States, Japan, and Europe.

- A durable U.S. expansion driven by small businesses, new investment, and inventory growth. Also, a glimmer of light on U.S. budget and demographic problems as population growth continues, the baby bust proves gradual, and economic growth generates extra tax receipts.
- Faster global growth than in the 1990s, enhanced by freer trade, globalization, and increasingly flexible economies.
- Another Asian miracle, this time driven by China, Japan, and India rather than by the "Tigers." The first Asian miracle, extending through the 1980s and 1990s, provided a huge lift to per capita incomes. I think another one is at work, triggered by China's dollar peg in the early 1990s having launched steady 8 percent growth, Japan's 2003 exit from deflation, and the accumulation of India's structural reforms over the last decade, one of the many dividends from the collapse of the communist economic model in 1989.

FIGURE 3.1 U.S. Real GDP Growth

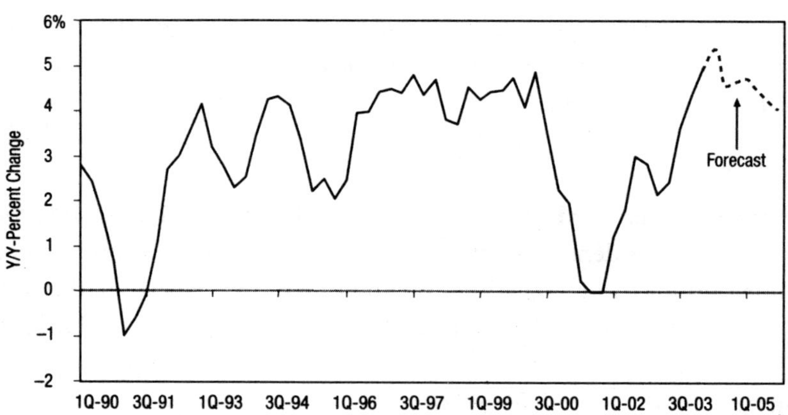

Source: Bloomberg; Bear Stearns & Co. Inc.

Currency Outlook a Key Variable

The 1990s expansion stopped, in my view, because the dollar got so strong it caused deflation in the United States and in dollar-linked countries. This contributed to the Asia crisis and worsened Japan's deflation spiral. High U.S. interest rates and tax rates, the strong-dollar-related U.S. investment extremes of the late 1990s, and super-low commodity prices helped end the expansion. In addition, oil, gasoline, and natural gas were at their most expensive in history. Adjusting for the change in the dollar's value, oil was 7.2 barrels per gold ounce in September 2000, versus "only" 11 at the end of 2004.

With those late-1990s problems resolved, the default option for the global economy is moderate-to-strong growth. I disagree with the view that growth is fragile.

The key variables in the strength of the expansion are terrorism, U.S. inflation, "animal spirits" (the degree to which corporations use their cash-flush balance sheets), and especially the value of currencies relative to their long-term averages.

After a bout with deflationary pressures in 1997–2001, the dollar's value is in a broad range that points to moderate inflation and a long global expansion. Global growth would be even faster if the dollar were more stable.

The principal dynamic at work in world currency markets—rather than the U.S. trade deficit, China's economy, or a foreign central bank—is the economic strength in the United States and the balance between low interest rates and high levels of deflation-related liquidity. I expect the U.S. to raise rates faster than current market expectations. If so, this would offset the excess liquidity, limit further commodity gains, and maintain the value of the dollar.

FIGURE 3.2 U.S. Dollar Price of Gold, 1970–2002

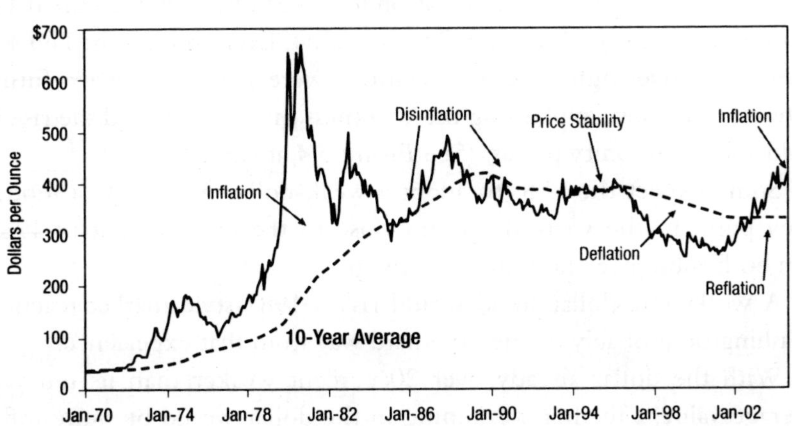

Source: Bloomberg; Bear Stearns & Co. Inc.

Commodity averages have showed renewed strength, reflecting the view that the world is growing fast and the Fed won't react. The CRB commodity index closed on October 1, 2004 at its highest since 1981. Gold reached $455 per ounce—its highest since the 1987 dollar/stock market crash. (See ***Figure 3.2*** on the previous page.)

Our index of industrial material prices excluding oil has been fluctuating near its all-time high hit on March 23, 2004. (This index of commodity prices did not increase in the 1970s nearly as much as did other indices, reflecting industrial prices rather than speculative precious metals and oil.)

FIGURE 3.3 **Industrial Materials Prices (excluding Oil), 1997–2004**

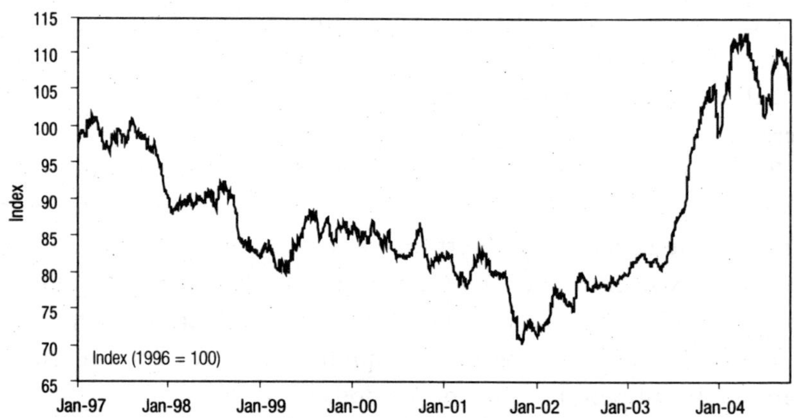

Source: Haver; Bear Stearns & Co. Inc.

The U.S. interest rate expectation for December 2005 declined 115 basis points between June and October 2004. Even though the Fed was raising the overnight rate, expectations were going down for future interest rate hikes, explaining the weakness in the dollar and the rise in gold and commodity prices. (See ***Figure 3.4*** at right.)

Some expect the U.S. to adopt a weak-dollar policy. As a model, they point to the weak-dollar response to the trade deficit in 1985. On both substance and politics, I disagree.

- A weakening dollar trend would risk a 1987-style market reaction. Washington probably learned the hard way from that experience.
- With the dollar already over 20 percent weaker than its ten-year average value, a further weakening in the dollar would be quite inflationary.

FIGURE 3.4 **Yield Implied from December 2005 3-Month Eurodollar Futures Contract**

Source: Bloomberg; Bear Stearns & Co. Inc.

• The history of currency fluctuations offers little evidence that a weaker dollar would correct the much-discussed imbalances. A trend toward dollar weakness would actually reduce U.S. savings, undercutting one of the goals of weak-dollar advocates.

Net foreign purchases of long-term securities have continued to substantially exceed the current account deficit, contradicting one of the key bearish concerns about the dollar and the trade deficit.

FIGURE 3.5 **Net Foreign Purchases of Long-Term Securities and the Current Account Deficit**

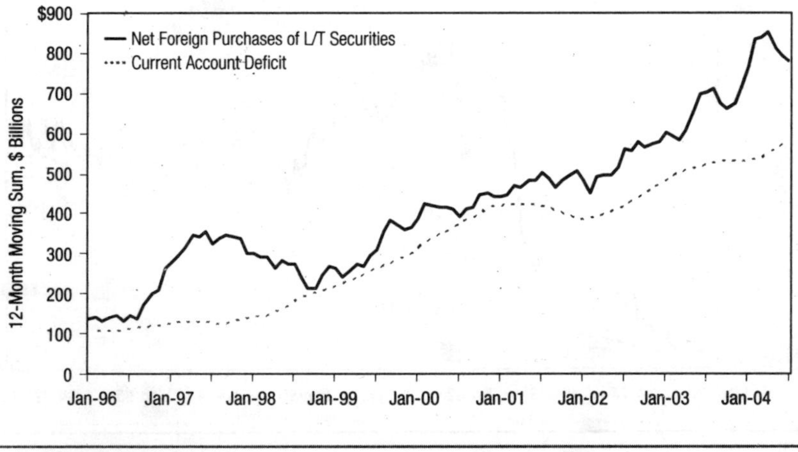

Source: Haver Analytics; Bear Stearns & Co. Inc.

The euro–dollar relationship should fluctuate less going forward than it did in the early years of the euro. True, the Eurozone runs a trade surplus and the United States a deficit. However, the stronger currency dynamic is the U.S. growth rate and expectations of relatively rapid U.S. rate hikes.

FIGURE 3.6 U.S. Dollars per Euro

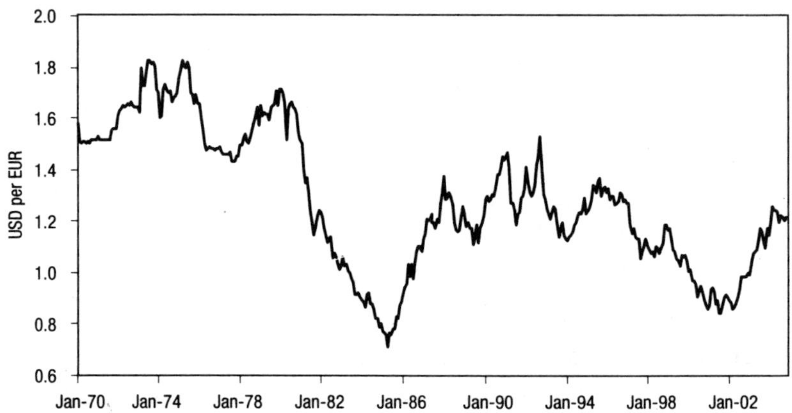

Source: Haver Analytics; Bear Stearns & Co. Inc.

Whereas the dollar moved substantially stronger than its ten-year moving average in the late 1990s, causing deflation, the euro's value remained relatively stable, as did its interest and inflation rates.

FIGURE 3.7 Euro Price of Gold

Source: Haver Analytics; Bloomberg; Bear Stearns & Co. Inc.

The yen is now valued on the view that Japan's economy is not gaining much strength and that interest rates will remain at zero percent for a long time. I disagree. The yen should benefit from Japan's exit from deflation, continued fast growth in China, and the strength of the U.S. economy. Those factors should tend to strengthen the yen versus both gold and the dollar.

FIGURE 3.8 **Yen Price of Gold**

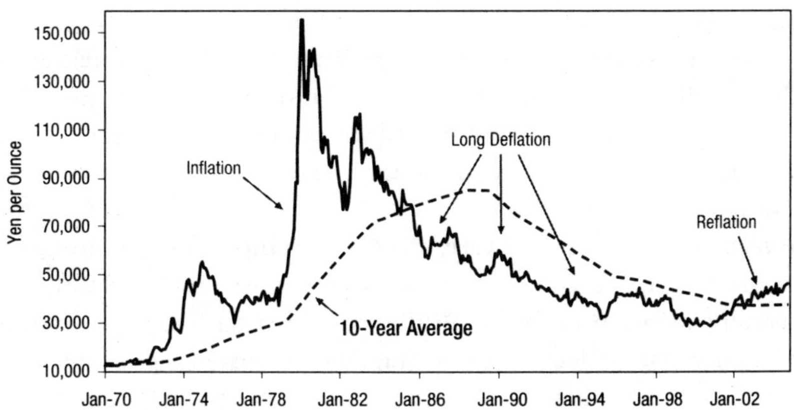

Source: Haver Analytics; Bloomberg; Bear Stearns & Co. Inc.

FIGURE 3.9 **Japanese Yen per U.S. Dollar**

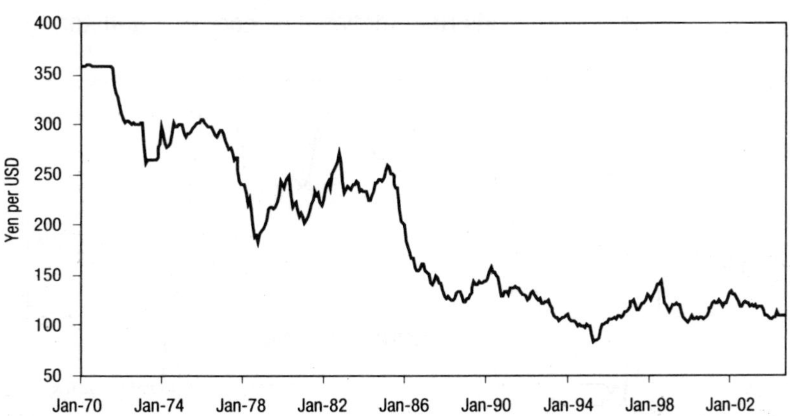

Source: Haver Analytics; Bear Stearns & Co. Inc.

Near-Term Outlook

The United States is likely to enjoy solid growth into 2005, answering once again the perennial forecasts of a substantial economic slowdown. Recall the slowdown discussion in September 2003 relating to slower mortgage refinancings and the expiration of the tax rebates. Another slowdown discussion occurred in June 2004 relating to excess auto capacity and weakness in the payroll survey. The reality has been fast growth rates driven by consumer resilience and an increasingly expansionary business sector.

The foundations for continued growth are in place: a reasonably valued dollar (we think somewhat inflationary), strong growth in developing countries, robust small-business profits, a recent U.S. tax cut with back-loaded growth benefits, and a 5.5 percent unemployment rate.

We are probably in the early stages of a multiyear expansion, characterized by strong U.S. and global growth, rising interest rates, and a persistent inflation problem. While expensive oil, corporate caution, and equity market declines cause drag, the expansion is probably strong enough to outlast them.

Real interest rates are negative, indicating a loose monetary policy. In effect, the Fed has begun to remove pressure from the accelerator but hasn't yet touched the brake.

FIGURE 3.10 **Real Fed Funds Rate (deflated by core PCE deflator)**

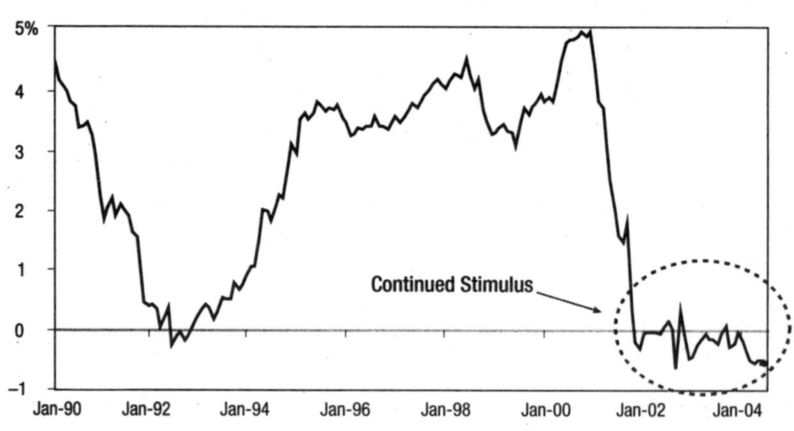

Source: Haver Analytics; Bear Stearns & Co., Inc.

The monetary stimulus is global. Overnight interest rates are at only 2.25 percent in the fast-growing United States, 0 percent in Japan, and 2 percent in Europe, versus the 4.75 percent benchmark rate in England. China's monetary policy is loosely linked to the expansionary U.S. monetary policy through its currency peg, adding further stimulus. This is not the stuff of slowdowns.

I also note huge corporate profitability and cash accumulation. U.S. corporate earnings climbed to a major record in the fourth quarter of 2003. Annualized NIPA-based U.S. corporate earnings in the fourth quarter were $1.2 trillion, up 29 percent year over year. Full-year 2003 corporate earnings were $1.1 trillion, 18 percent above 2002, the previous record.

FIGURE 3.11 **U.S. Corporate Profits (NIPA)**

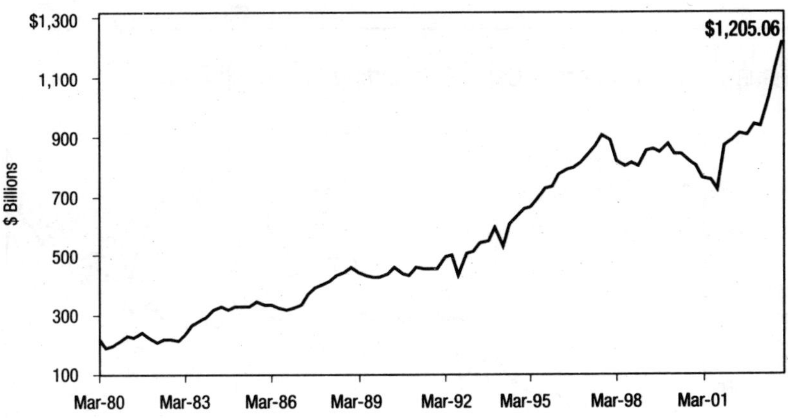

Source: Haver Analytics; Bear Stearns & Co., Inc.

From the second quarter of 2003 to the first quarter of 2004, corporations had been in the odd position of hoarding cash flow. The corporate financing gap returned to positive territory in the second quarter of 2004, meaning corporations spent more in the second quarter than their cash flow. (See *Figure 3.12* on the following page.)

In the second quarter of 2004, corporate cash flow increased by $22 billion, while corporate capital spending increased by $48 billion. Economic growth can be expected to shift increasingly toward the corporate sector. (See *Figure 3.13* on the following page.)

FIGURE 3.12 **Corporate Financing Gap**

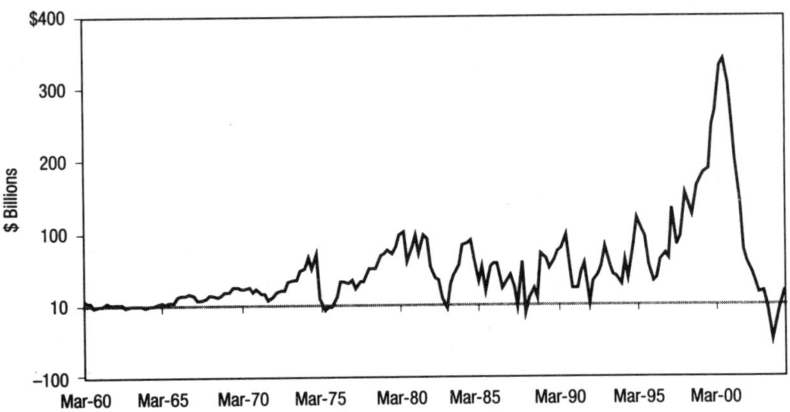

Source: Haver Analytics; Bear Stearns & Co., Inc.

FIGURE 3.13 **Corporate Capital Spending and Cash Flow**

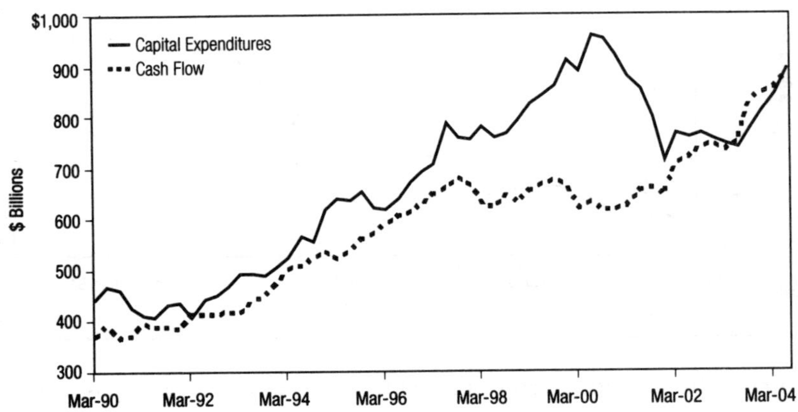

Source: Haver Analytics; Bear Stearns & Co., Inc.

There was extensive coverage of the economy's "slowdown" or mid-2004 soft patch. Rather than constituting a trend, the slowdown was caused by several one-time factors, in particular the nationwide closures for President Reagan's funeral, the second-quarter gasoline price spike, the cool weather, and, importantly, the downswing in the auto sector in the second quarter as it struggled to reduce incentives.

One of the remarkable aspects of the 2003–2004 expansion was the decline in inventory and the inventory/sales ratio. The ratio fell to an all-time low. While this held back growth and employment, it also suggests a simple source for future growth—the process of rebuilding inventory. Even a reversion to the secular downtrend would add almost 0.9 percent to a full-year's GDP, assuming trend growth in sales.

FIGURE 3.14 **Ratio of Inventories to Sales, 1968–2004**

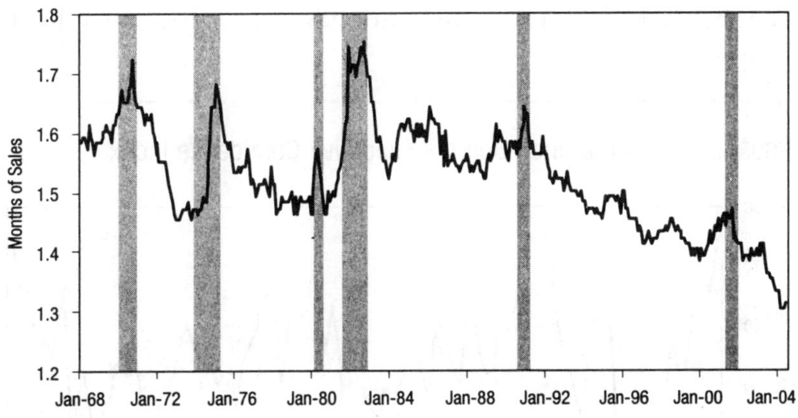

Source: Haver Analytics; Bear Stearns & Co., Inc.

Manufacturing is another indicator pointing to economic strength rather than weakness. In 2003 and the first part of 2004, the Institute for Supply Management (ISM) index stayed above 60 for nine consecutive months—for the first time in twenty years.

FIGURE 3.15 **ISM Index**

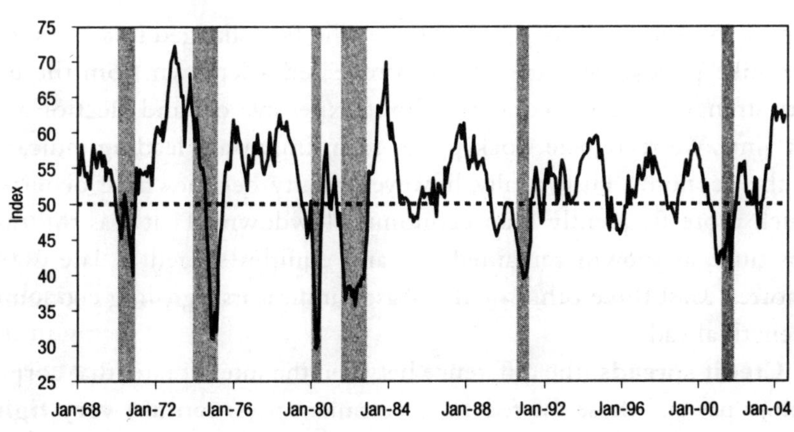

Source: Haver Analytics; Bear Stearns & Co., Inc.

According to the ISM, the current level of its index is consistent with GDP growth above 5 percent. Manufacturing job growth has strengthened substantially.

One explanation for weak growth in inventories and employment after the 2001 recession was the risk-aversion and caution of U.S. CEOs. Their confidence fell to the lowest levels since the "malaise" of the late 1970s. In 2004, CEOs became significantly more optimistic, helping explain the strong growth and pointing to a durable expansion.

FIGURE 3.16 **Conference Board's Executive Confidence Index**

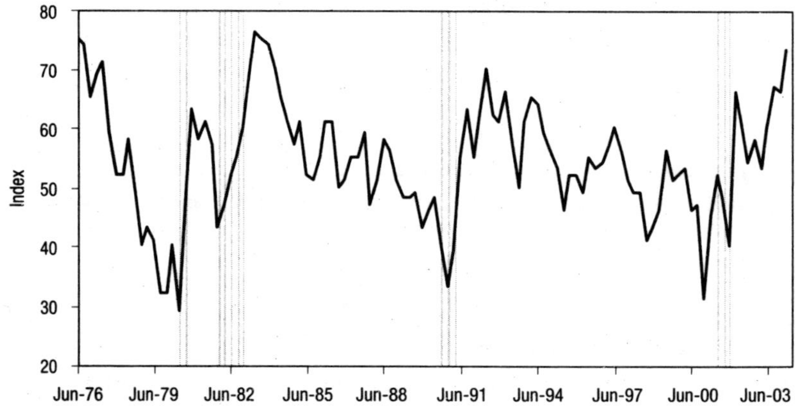

Source: Bloomberg; Bear Stearns & Co., Inc.

Market-Based Indicators Signaling Strong Growth

Equity markets weakened in 2004 after the Fed changed to a measured rate hike policy. We think this also reflected a letdown from the equity strength in 2003, concerns about expensive oil, and election uncertainty. We recognize stock prices as an important leading indicator of the economy. Historically, however, equity declines have occurred much more frequently than economic slowdowns. That was the case this time, as growth remained fast and equities surged in late 2004. I note at least three other market-based indicators signaling economic strength ahead:

- **Credit spreads** (the difference between the interest rates that corporations pay and those the federal government pays) **remain very tight**, signaling a strong economy rather than a weak one.

FIGURE 3.17 **BBB Yield Minus 10-Year Treasury Yield**

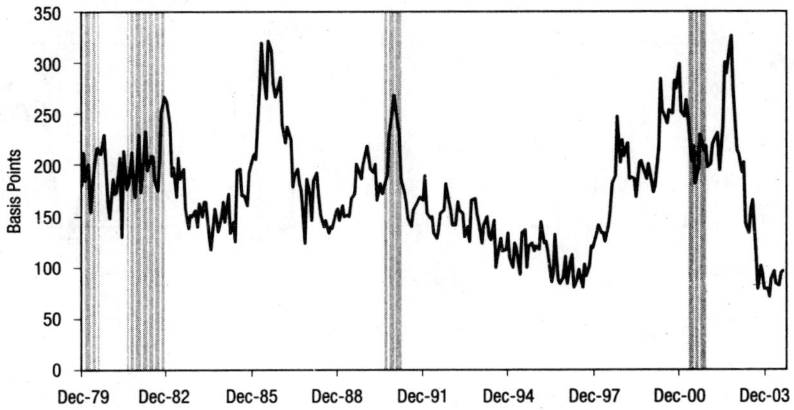

Source: Haver Analytics; Bear Stearns & Co., Inc.

FIGURE 3.18 **Industrial Materials Prices (excluding Petroleum)**

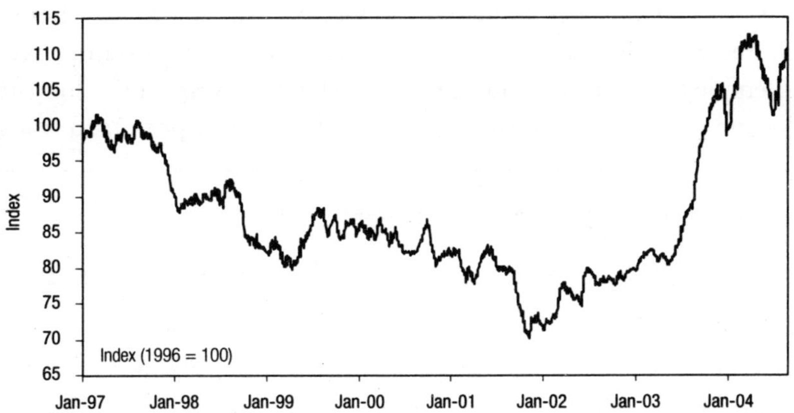

Source: Haver Analytics; Bear Stearns & Co., Inc.

- **Industrial materials' prices remain high**, reflecting the stimulative monetary policy and strong global growth.
- **A steep yield curve** (the difference between a long-term interest rates and short-term interest rates) **usually signals economic strength** due to monetary stimulus and the market-based expectation of higher future yields (and interest rates) than current yields. (See *Figure 3.19* on the following page.)

FIGURE 3.19 **Yield Curve (10-year Treasury yield minus 1-year Treasury yield)**

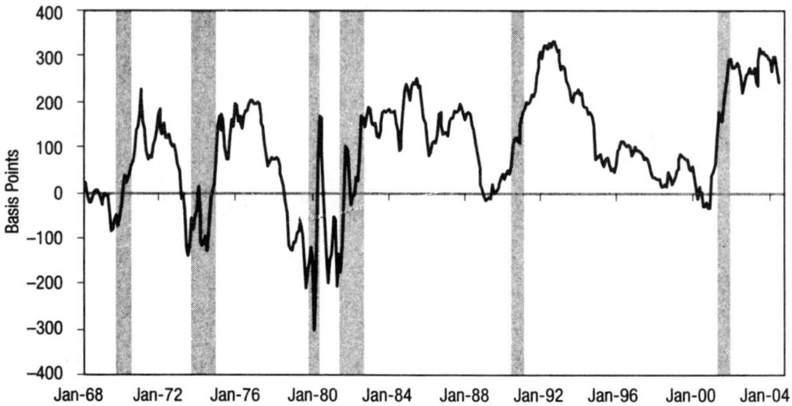

Source: Haver Analytics; Bear Stearns & Co., Inc.

Labor Issues

One of the driving influences in longer-term optimism about the United States is the flexibility of the labor force and its attractiveness to employers from around the world. This shows up in the long-term decline in U.S. unemployment, with steadily lower peaks and troughs.

FIGURE 3.20 **U.S. Unemployment Rate**

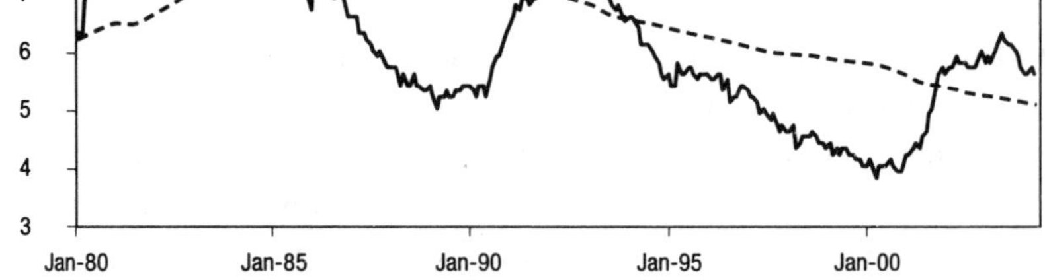

Source: Haver Analytics; Bear Stearns & Co., Inc.

The labor force participation rate (the portion of the over-16 civilian population that is employed or looking for a job) has reached high levels over the last decade, pointing to a well-trained work force with bright prospects for the future.

FIGURE 3.21 **U.S. Labor Force Participation Rate, 1960–2000**

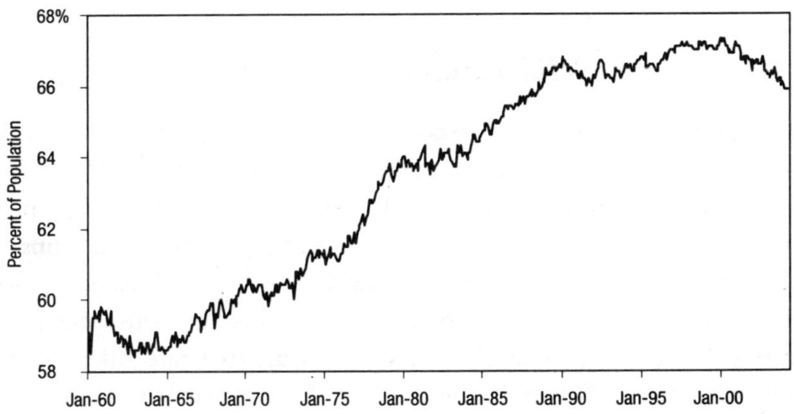

Source: Haver Analytics; Bear Stearns & Co., Inc.

FIGURE 3.22 **Manufacturing Employment as a Percentage of Total, 1954–2003**

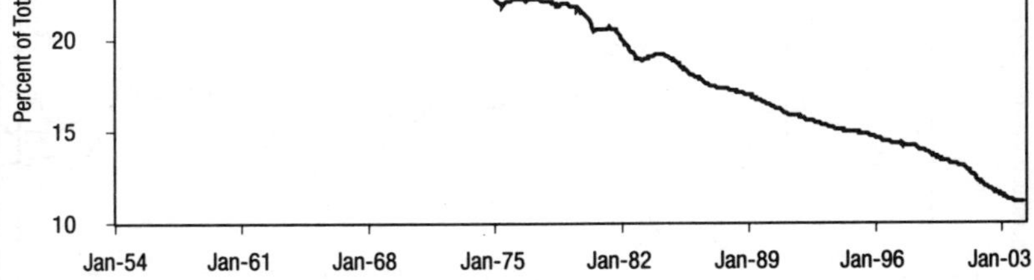

Source: Haver Analytics; Bear Stearns & Co., Inc.

The character of the U.S. labor force went through massive changes in the twentieth century, shifting from agriculture to manufacturing to services. Manufacturing employment as a share of total employment has been declining for more than fifty years. This is not a phenomenon specific to this business cycle or even to the United States. It has occurred in every large industrial country. (See ***Figure 3.22*** on the previous page.)

Solid Household Finances

There has been a big misunderstanding about the consumer's health in recent years. Some argue that the consumer didn't save much. This is factually incorrect. Consumers benefited broadly from low unemployment and falling prices in the late 1990s. They saved a portion of the gains.

In a February 23, 2004 speech to the Credit Union National Association, Fed Chairman Greenspan explained that the consumer has a strong balance sheet (strongly implying, to us, that rate hikes would not be disruptive): "Overall, the household sector seems to be in good shape, and much of the apparent increase in the household sector's debt ratios over the past decade reflects factors that do not suggest increasing household financial stress. And, in fact, during the past two years, debt-service ratios have been stable."

The "personal savings rate" statistic severely understates consumer savings. If realized flows on equities, houses, and mortgage refinancings are

FIGURE 3.23 **Adjusted U.S. Savings Rate**

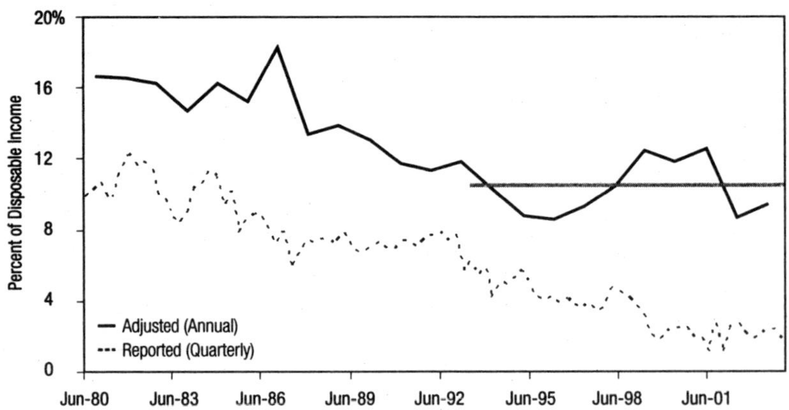

Source: Haver Analytics; CBO; Bear Stearns & Co. Inc. estimates.

included, the savings rate has been relatively stable over the last decade.

The Fed released its flow-of-funds data in September 2004. It confirmed the health of the U.S. household sector.

- Net worth had reached a new record (at $45.9 trillion), as had household assets and household debt. In addition, the household sector is liquid and is a big net creditor with a favorable maturity profile. Liquid assets (including direct equity and bond holdings, savings accounts and cash) reached $17.6 trillion, versus total liabilities of $10.1 trillion. The liabilities are dominated by mortgages ($7.1 trillion), of which roughly 80 percent are fixed rate over a long term. Thus, it is unlikely that the coming string of interest rate hikes will damage the household sector on net because it has both a favorable liquidity position and a favorable maturity advantage. We expect consumption to grow steadily into 2005, as the expansion proceeds.

FIGURE 3.24 **Household Balance Sheet**

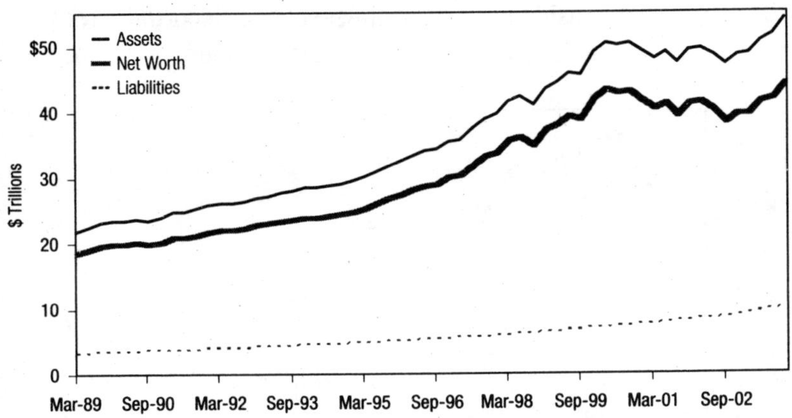

Source: Haver Analytics; Bear Stearns & Co. Inc.

- The Fed's flow-of-funds data showed that households increased their net additions to financial saving (this doesn't count real estate or stock market gains) to $852 billion at an annual rate in the second quarter of 2004. (See ***Figure 3.25*** on the following page.)
- In percentage terms, the savings rate indicated by the Fed's flow of funds was roughly 9 percent in the second quarter, versus the 0.6 percent recorded in July's personal savings data. (See ***Figure 3.26*** on the following page.)

FIGURE 3.25 **Net Acquisition of Financial Assets versus Reported Savings**

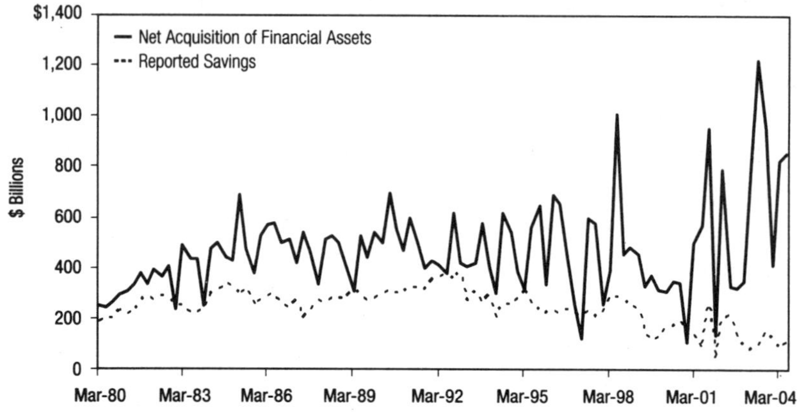

Source: Haver Analytics; Bear Stearns & Co., Inc.

FIGURE 3.26 **Acquisition of Financial Assets and Reported Savings as a Percentage of Disposable Income**

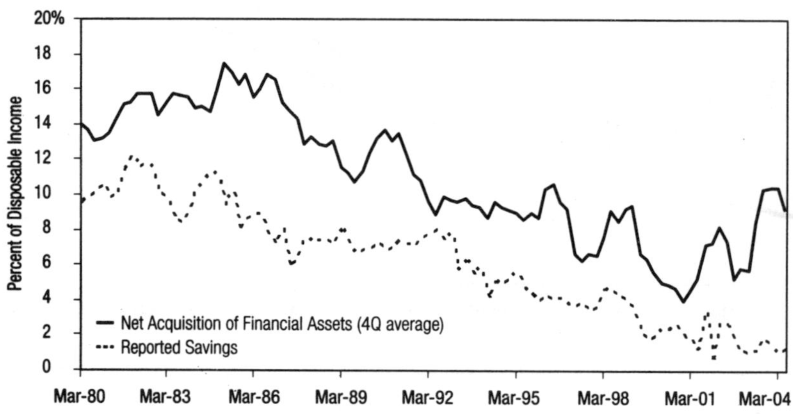

Source: Haver Analytics; Bear Stearns & Co. Inc.

The Housing Outlook

The expansion faces many obstacles and drags, including expensive oil, terrorism, capacity constraints in an increasing number of sectors, and past under-investment in many parts of the world. But I disagree with many of the fragile-recovery arguments, including a U.S. housing crash. Fed Chairman Greenspan has noted that bubbles are difficult to identify.

I agree, and add that in the case of housing, unlike equities or commodities, markets are heavily influenced by local factors, creating local excesses.

- The year 2004 saw record sales of new and existing homes.

FIGURE 3.27 U.S. New Home Sales (annual rate), 1970–2002

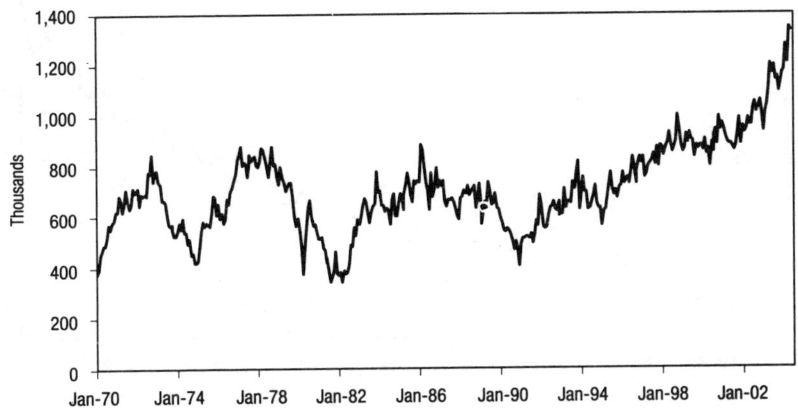

Source: Haver Analytics; Bear Stearns & Co. Inc.

- Some of the increase in new home sales reflects the increasing U.S. population. On a per capita basis, the rate of new home sales has been steadier (see *Figure 3.28* below). Adjusted for the obsolescence and replacement of old homes, net new homes per capita would be consistent with historical norms.

FIGURE 3.28 U.S. New Home Sales per Capita

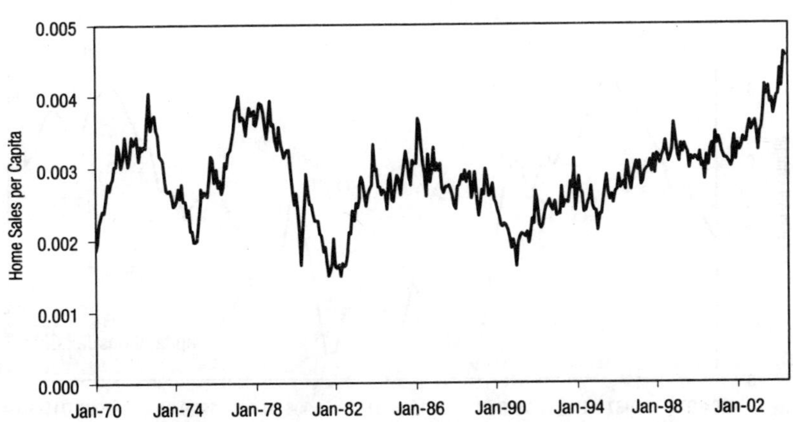

Source: Haver Analytics; Bear Stearns & Co. Inc.

• Along with high sales volume, house prices also have hit records. This is seen by some analysts as a bubble.

FIGURE 3.29 **U.S. Median Home Price**

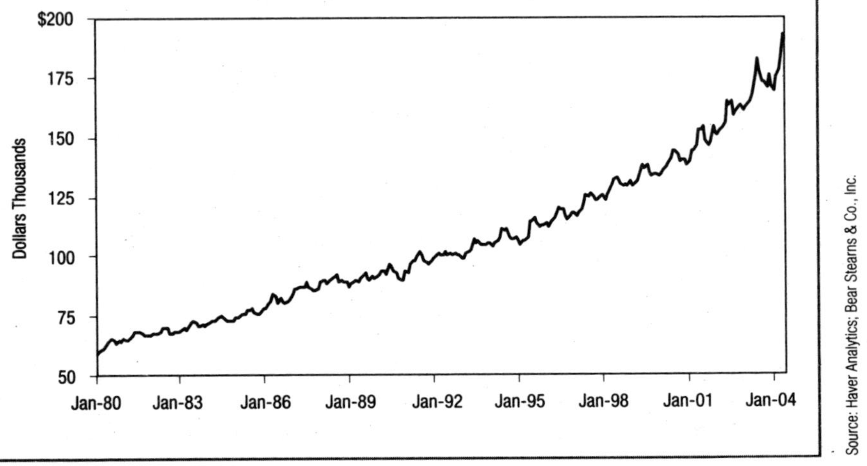

At the national level, there appear to be sound fundamental factors causing much of the current rise in house prices. Of course, speculation is also a factor, especially in some local areas.

• The 1997 cut in the capital gains tax rate on houses added substantially to the value of houses, just as the 2003 cut in the capital gains tax rate on equities increased the value of equities.

FIGURE 3.30 **U.S. House Prices**

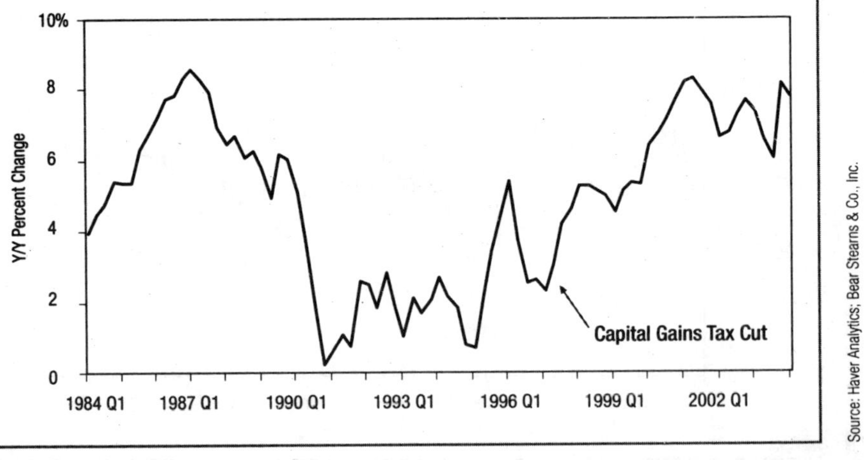

• House price increases had lagged the economy and the equity market in the mid-1990s, so some of the gains were a catch-up process. (See *Figure 3.30* at left.)

• Strong demographics are playing a critical role. Increased immigration and the formation of households by the children of baby boomers are providing a firm foundation for increased housing demand.

FIGURE 3.31 **Household Formation 29 Years and Younger**

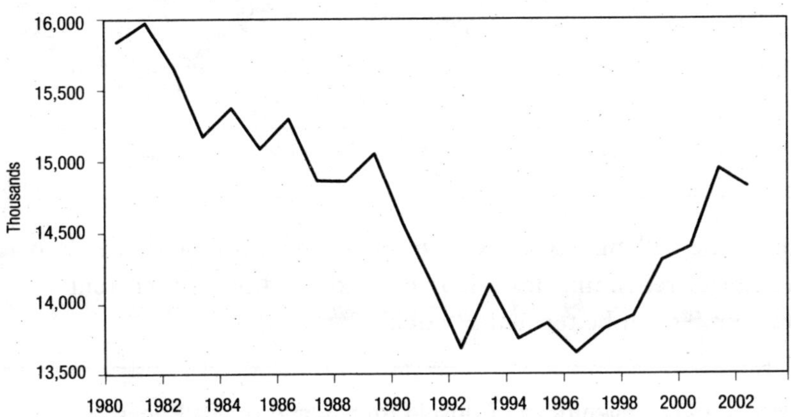

Source: Haver Analytics; Bear Stearns & Co. Inc.

• The long-term decline in the U.S. unemployment rate has added to the value of land and houses in the United States. Similarly, the decline in the UK unemployment rate and its avoidance of recession in 2001 added to the value of land and houses there, partially defending its even stronger house price gains in the face of high interest rates and mostly floating-rate mortgages. (See *Figure 3.32* on the following page.)

• Housing affordability also has provided a strong underpinning for the housing market. While house prices have increased, mortgage interest rates have been low and real disposable income growth has been solid.

• A key factor behind the rise in house prices has come from the supply side. In previous episodes of sharply rising house prices over the past 30 years, the increase in price has led to overbuilding. Once inventories of unsold homes reached a critical mass, prices tumbled. The current episode has been different. Home builders have not built heavily ahead of sales—inventories of unsold homes are fluctuating

FIGURE 3.32 U.S. Unemployment Rate

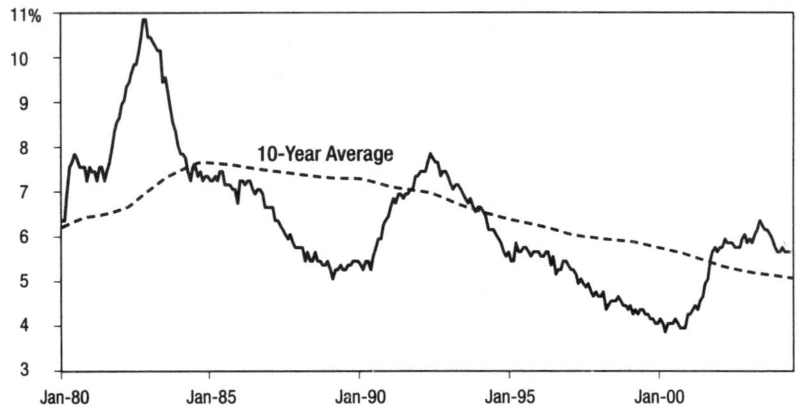

Source: Haver Analytics; Bear Stearns & Co. Inc.

around their all-time lows. With fundamentals pushing up demand and inventories remaining lean, it is hard to see how the current rise in house prices can be termed a "bubble."

FIGURE 3.33 Inventories of Unsold Homes and House Prices

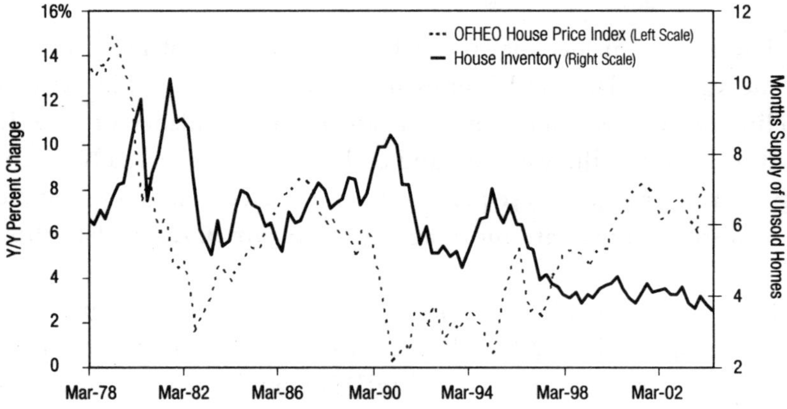

Source: Haver Analytics; Bear Stearns & Co., Inc.

I disagree with one of the premises of the housing bubble theory—that consumers have adopted risky practices in financing their homes which will destabilize the market if prices soften. The percentage of fixed-rate mortgages has remained at a high 80 percent level. (Due to fast turnover, an increasing number of new mortgages have had adjustable rates in response to the steep yield curve, but the stock of fixed rate mortgages has stayed relatively stable as mortgagers prepare for higher rates.) The aggregate equity in homes has also remained relatively stable at 55 percent of home values.

U.S. house prices may show localized weakness in coming quarters as mortgage rates rise. This would slow the gains in median home prices, but should not cause a decline in the nationwide median price. It would not be a crash, though some local areas could see sharp enough declines to be characterized as mini-bubbles. Separately, there will probably be some slowdown in new home sales and in the residential construction component of GDP as mortgage rates rise and affordability declines. Consumers may become satisfied with the quantity of their housing, shifting the focus of their consumption elsewhere.

FIGURE 3.34 **Square Footage of New Homes Sold**

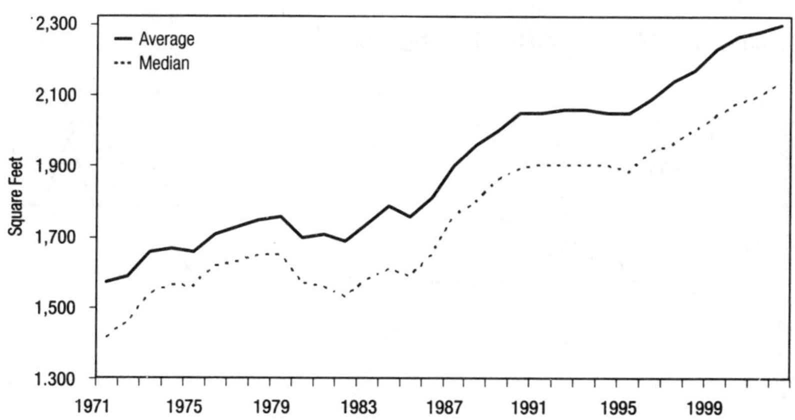

Source: Census Department; Bear Stearns & Co., Inc.

Global Expansion Underway

I expect a durable, multiyear global expansion driven by continued growth in U.S. capital investment, growth in foreign consumption, and a drawn-out global rebuilding of inventories.

The shift toward reflation, low U.S. real interest rates, and higher commodity prices creates a much better environment for developing countries than the 1980–2000 combination of disinflation, high real interest rates, and lower commodity prices. For example,

• There have been substantial improvements in Brazil, South Africa, Turkey, and Eastern Europe.

• In Asia, broad, multiyear gains in living standards and equity market capitalization are likely.

• China is continuing to grow fast, aided by currency stability, market liberalization, consumer gains, rural development, and foreign investment. Recent monetary and regulatory changes are constructive, adding to China's long-term growth potential.

• India is benefiting from growth in business investment and domestic demand sparked by past structural reforms and increased credit penetration. We expect a muted impact from the erratic 2004 monsoons on farm output and consumption demand in rural areas.

• Industrial growth in Asia remains very strong, with double-digit growth recorded across the board. China's industrial production rose 15.7 percent year over year in October 2004 (15.9 percent three-month average).

FIGURE 3.35 **Asian Industrial Production**

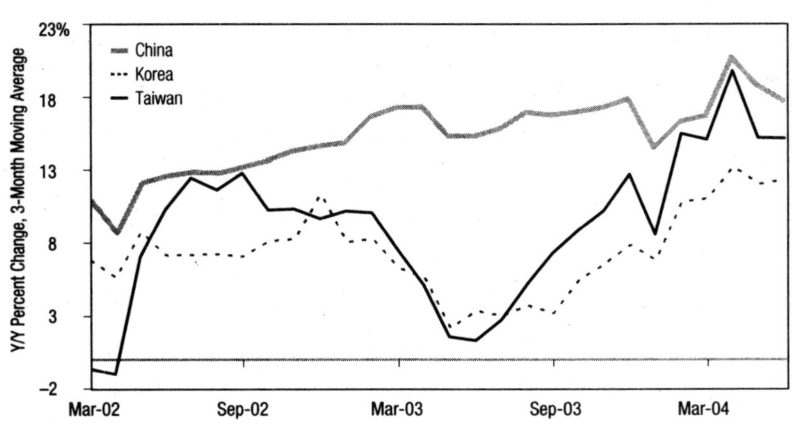

Source: Bloomberg; Bear Stearns & Co., Inc.

• Industrial output is surging in developing Europe, led by double-digit growth in Turkey and Poland and steady growth in Russia (7.8 percent in October, 9.4 percent three-month average). (See *Figure 3.36* at right.)

• While growth in Latin America is slow relative to Asia and the United States, an industrial rebound is underway. Output in Brazil has surged. (See *Figure 3.37* at right.)

FIGURE 3.36 **EMEA Industrial Production**

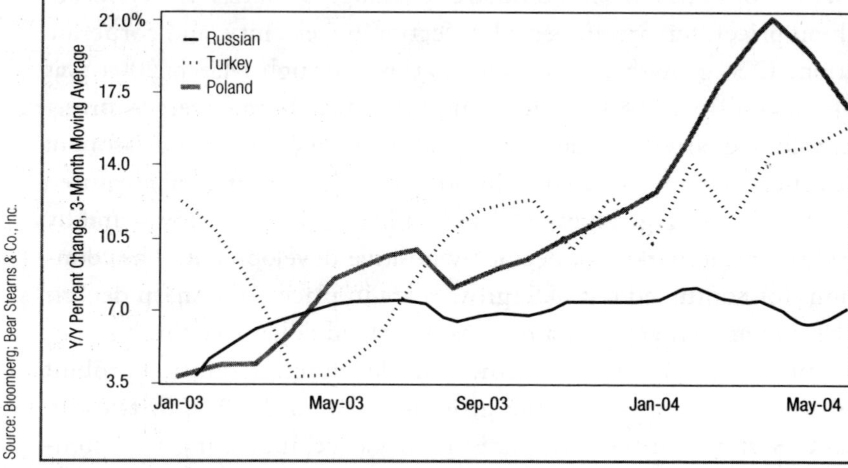

FIGURE 3.37 **Latin America Industrial Production**

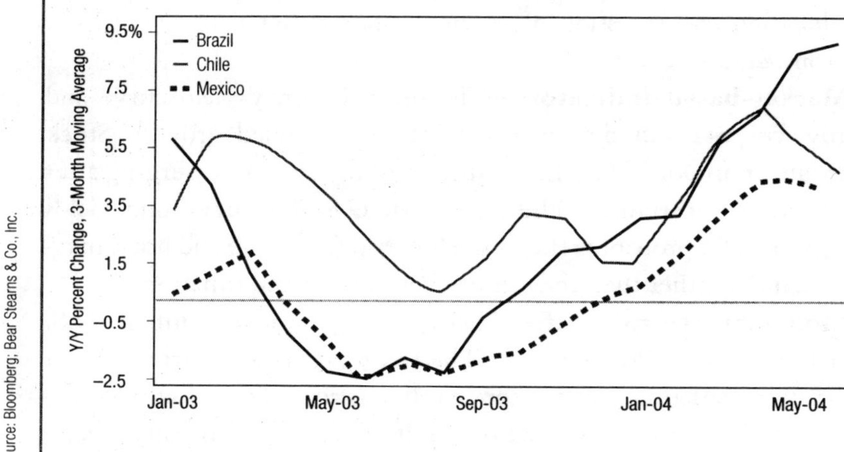

Likely Trends

To recap the message of the indicators we have discussed in this chapter, here are some of the developments we are likely to see as we move through the second half of this decade:

Durable expansion, moderate inflation—We are probably in the early stages of a multiyear expansion characterized by solid U.S.

and global growth, rising interest rates, and persistent inflation. This expansion was preceded by deflation, making it markedly different from previous ones. It should be strong enough to outlast the effects of high oil prices, terrorism, periodic election uncertainty, and corporate caution. U.S. growth in the four quarters through March 2004 was the strongest since 1984, the four quarters through June were as strong as any four quarters in the 1990s, and forward-looking indicators of production remain elevated. The idea of a long string of measured rate hikes is being grudgingly absorbed by the U.S. economy and by world financial markets, a decidedly positive development. The foundations for continued fast U.S. growth are in place—the sharp decline in the unemployment rate, a reasonably valued dollar (we think somewhat inflationary), stronger growth in developing countries, robust small-business profits, and the growth-oriented 2003 U.S. tax cut. Real U.S. short-term interest rates are negative, indicating an accommodative monetary policy. Japan's interest rate is still zero percent. There is huge corporate and household liquidity in the United States and Japan. Meanwhile, U.S. and foreign inventories are low, reflecting the deflation and recession of previous years rather than the current environment.

Market-based indicators—The steep Treasury yield curve and narrow corporate credit spreads point to continued growth. Stock prices are an important leading indicator of the economy, surging after the election uncertainty. The decline in U.S. Treasury bond yields since June 2004 probably reflects the low Fed funds rate and bond market dynamics rather than the outlook for growth and inflation.

Monetary regime shift—There has been a major shift in U.S. monetary policy following 9/11. The twenty-year-long strong-dollar disinflation/deflation process stopped. It was replaced by lower nominal and real interest rates, a weaker dollar, higher commodity prices, improved prospects for growth outside the United States, and Japan's exit from deflation. Interest rates in the United States, Europe, Japan, and China will likely rise toward neutral in a multiyear, somewhat inflationary process, the inverse of the disinflation of the 1990s. Bond yields should seesaw upward in coming years, continuing the pattern started in 2003.

Global demand and profit growth—This will probably bring a shift toward faster global demand growth. In the United States, the focus of demand growth is evolving constructively toward business investment, inventory rebuilding, and job growth. There are signs of pent-up consumer demand in much of the world outside the U.S. due to recent deflation and recession. In the Eurozone, domestic demand growth should spread beyond France, rising toward "half-the-U.S." growth-rate expectations. Despite Japan's disappointing GDP growth rate, we think it is over the hurdle in terms of exiting its deflation. World nominal dollar GDP should reach $36.6 trillion in 2004, up 10.7 percent in 2004, and 11.6 percent in 2003. This provides a strong platform for corporate dollar profits.

In 2002 and 2003, growth and the exchange rate were key variables in the strength of the expansion. In this chapter, I presented some of the data used in explaining the evolution of the U.S. and global economies. While the data will change going forward, I think the same analytical techniques will apply through 2005 and beyond. Key variables are likely to be inflation and the exchange rate.

4

How U.S. Fiscal Policy Aids Both U.S. and Global Growth

KATHLEEN STEPHANSEN

KATHLEEN STEPHANSEN presents a concise chapter, no surprise to her legion of fans. As director of global economics for Credit Suisse First Boston, Stephansen is known for an economy of words, a closeness of thought that borders on poetry. Here, Stephansen creates a set piece on U.S. fiscal economics and how intertwined our balance sheet is with other fiscal experiments. Years ago, her fiscal analysis was found in Donaldson, Lufkin & Jenrette's weekly "Maroon Book." It made for required reading. So too, this chapter.

U.S. fiscal policy over the past three years has been instrumental in buffering the cyclical downturn, not just in the United States but also in the global economy. Increasingly, with rising economic integration, policy changes in one country have had a large impact on economic activity in other countries. Furthermore, the U.S. economy's high propensity to consume and to import, combined with the country's policy flexibility, has reaffirmed the United States as the locomotive to global growth.

By mid-2004, with global growth close to a thirty-year high, calls for the fiscal pendulum to swing toward more restraint have gathered momentum. But the return to fiscal discipline appears uncertain and, with Social Security and tax reform on the agenda for the next couple of years, a countercyclical fiscal policy is in the offing—that is, a still-expansionary fiscal policy in an economy that has reached (or is close to reaching) full employment (see ***Figure 4.1*** on the following page).

FIGURE 4.1 **Fiscal Policy Stance, 2001–2005 (Projected)**

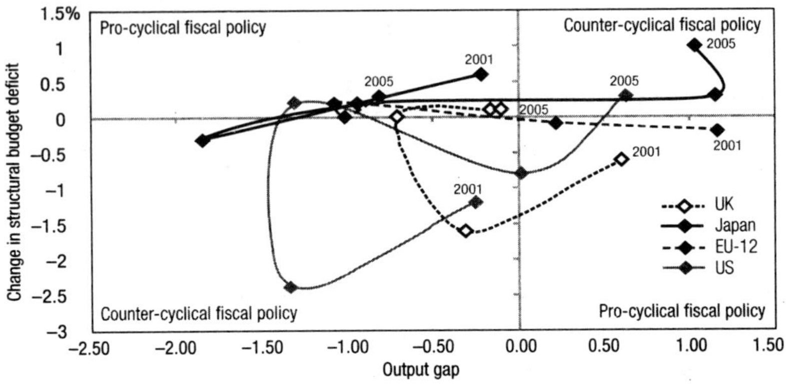

Source: Datastream International, OECD, Credit Suisse First Boston (CSFB)

The return to fiscal discipline, though highly desirable at this juncture of the business cycle, will likely be slow. Continuous foreign capital inflows play an important role in financing the fiscal imbalance and help postpone hard policy choices.

This chapter discusses how changes in the budget profile are likely to affect economic growth, both in the United States and the global economies. It also discusses the current state of fiscal policy.

Changes in the Budget Profile

The U.S. fiscal position has changed dramatically in the past four years. In 2000, surpluses were projected over the next decade. By 2002, plusses turned to minuses, and the budget balance swung from a $236 billion surplus in 2000 to a $377 billion deficit in 2003. The cumulative deterioration totaled a whopping $613 billion. In GDP terms, this represents 5.9 percentage points (see ***Figure 4.2***).

Of the deterioration, $180 billion reflects "cyclical" factors, that is, changes in the business cycle, and $420 billion reflects "legislative/structural" factors, such as the change in the tax law and enacted additional spending. Another $13 billion is attributed to "technical" adjustments. Revenues were the hardest hit, with their decline amounting to 75 percent of the budget's total deterioration. Outlays rose by 1.5 percentage points over that period, representing 25 percent of the budgetary deterioration.

FIGURE 4.2 **Budget, Revenues, Outlays (as % of GDP)**

	2000	2003	Change (%)	Share of Change (%)
Budget Surplus (+)/Deficit (-)	**2.4**	**-3.5**	**-5.9**	**100**
Revenues	**20.9**	**16.5**	**-4.4**	**75**
Individual	10.3	7.3	-3	51
Corporate	2.1	1.2	-0.9	15
Outlays	**18.4**	**19.9**	**1.5**	**25**
Discretionary	6.3	7.6	1.3	22
Defense	3	3.7	0.7	12
Domestic	3.1	3.6	0.5	8
Entitlements	9.8	10.9	1.1	19
Social Security	4.2	4.3	0.1	2
Medicare	2.2	2.5	0.3	5
Medicaid	1.2	1.5	0.3	5
Income Support	1.4	1.8	0.4	7
Interest	2.3	1.4	-0.9	-15

Source: U.S. Treasury; CSFB

Three notable factors causing the swing from surplus to deficit are worth highlighting—one cyclical and two legislative.

Cyclical Developments

The first factor pertains to the effects of the business cycle on tax revenue and spending growth. The 2001 recession and subpar 2002 recovery generated a contraction in overall tax receipts in fiscal year 2001 (FY01) and FY02. Businesses were hardest hit during the recession, and the weakness in corporate profits, and hence corporate tax receipts, was the biggest factor holding down revenues in FY01. Weak individual tax receipts accounted for virtually the entire shortfall in revenues in the following two years (FY02 and FY03), due to:

- a lack of capital gains—2001 stock market losses realized on tax returns in April 2002 and 2003 drove down *not withheld* income taxes; and
- the weakest postwar labor market recovery, which depressed *withheld* income taxes.

Spending also is affected by the cycle, notably in the form of higher spending on unemployment benefits. As interest rates fell, the higher spending was offset, at least partially, by lower debt-servicing costs.

Legislative Changes

Legislative changes over the past three years were by far the more important factor having contributed to the shift from budget surpluses to deficits. For a starter, FY02 began just weeks after 9/11, and defense outlays rose sharply as the war on terrorism began. War financing implies waiving, at least temporarily, any mechanisms for containing spending and efforts to resolve longer-term issues such as Social Security funding. Defense spending rose from 3 percent to 3.7 percent of GDP between FY00 and FY03, and non-defense discretionary spending rose, albeit by a slightly tamer 0.6 percentage point.

Then, three ten-year tax cut packages were enacted: The $1.348 trillion tax cut package passed in 2001, which provided households with a $40 billion tax rebate; the $42 billion Jobs Creation and Worker Assistance Act of 2002; and the $350 billion Jobs and Growth Tax Relief Reconciliation Act of 2003. The latter comprises measures that accelerate the tax cuts enacted in the 2001 package (costing an estimated $171 billion over ten years); growth incentives for businesses, such as more generous depreciation incentives (costing an estimated $10 billion over ten years); and reductions in taxes on capital gains and dividends (costing an estimated $148 billion over ten years).

The Cyclical versus the Structural Budget

The budget deficit should continue to widen. The deficit rose to $413 billion in 2004, up from 2003's $377 billion shortfall. But the economic recovery reduced the cyclical deficit considerably, from $90 billion in 2003 to close to $35 billion in 2004. Therefore, the bulk of the imbalance is now structural. Of 2003's $377 billion deficit, the Congressional Budget Office calculates the structural deficit to have totaled $310 billion (once the cyclical budget and technical adjustments have been taken into account), the highest structural deficit level ever and well above the prior peak of $211 billion reached in 1986. For 2004, the structural deficit reached $330 billion. As a share of potential GDP the profile is less dire: 2003's structural deficit stood at 2.8 percent and remained close to that share in the fiscal year 2004, but down significantly from the 4.8 percent

FIGURE 4.3 **Budget Deficits (% of GDP, by fiscal years), 2000–2014**

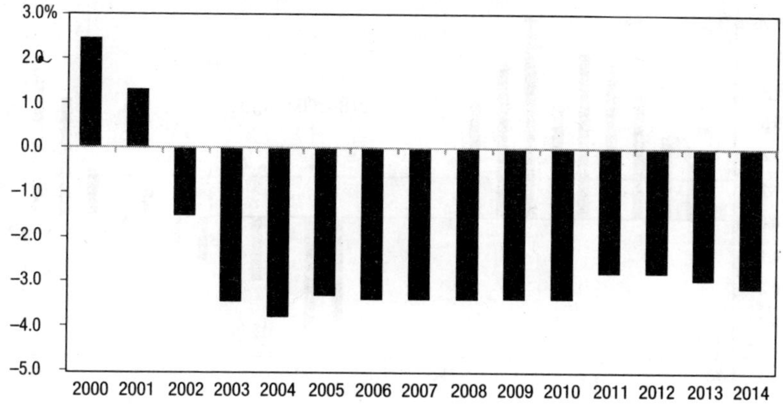

Source: U.S. Treasury; CSFB

share in 1986. This being said, the dynamics are clear: Economic growth will not be sufficient to achieve any meaningful correction in the deficit. It will necessitate legislative measures in the form of tax increases and/or spending cuts. Without any legislative change and assuming steady economic growth during the remainder of the decade, the total budget (cyclical + structural) will continue to post sizable deficits on the order of 3.5 percent as a share of GDP (see ***Figure 4.3***).

The Effects on the Economy§

The countercyclical U.S. fiscal impulse had tangible effects on the U.S. economy. In 2002 and 2003, both defense and non-defense spending had pulled more weight in GDP (see ***Figures 4.4*** and ***4.5*** on the following page), the aggregate tax burden had declined dramatically, while some rebalancing of the U.S. financial structure in favor of safer Treasuries and a less-leveraged private liability structure have helped corporate balance sheets recover.

Combined corporate tax payments and personal tax payments (including payroll and state/local taxes) as a share of nominal GDP stood at 17.6 percent in the third quarter of 2003, its lowest in almost

§ The discussion of the effects on the economy, specifically for the U.S. household and corporate sectors, draws on CSFB's U.S. economics department's analyses published in the CSFB *US Economics Digest*.

FIGURE 4.4 **National Defense Contribution to GDP***

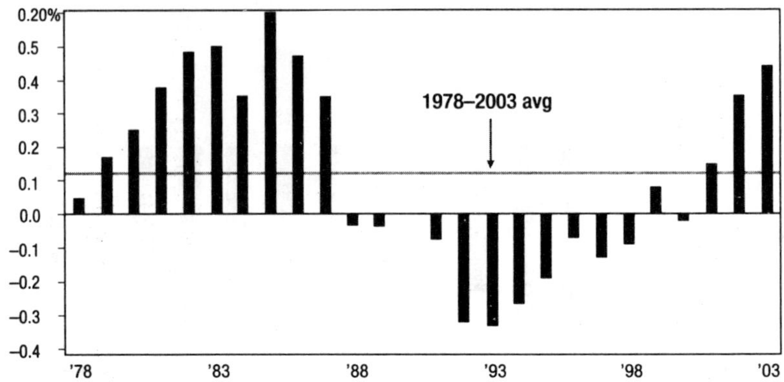

* Annual percentage point contribution to GDP growth

Sources: BEA; CSFB

FIGURE 4.5 **Non-Defense Contribution to GDP***

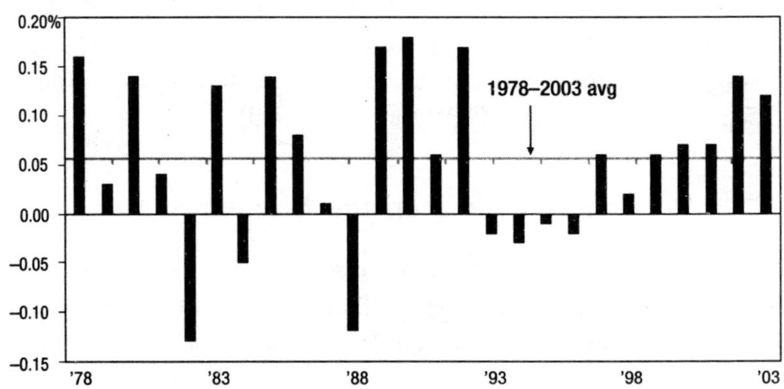

* Annual percentage point contribution to GDP growth

Sources: BEA; CSFB

thirty years, and a full five percentage points below the peak in the first quarter of 2000 (see ***Figure 4.6*** at right).

Part of the reason for the decline in the tax share over the past few years was the steep drop in capital gains taxes during the stock market downturn, which essentially reversed the massive build-up of capital gains taxes during the bull market in the late 1990s. Tax revenue growth also slowed because there were fewer individual taxpayers (rising unemployment) and

FIGURE 4.6 **Personal and Corporate Tax Payments***

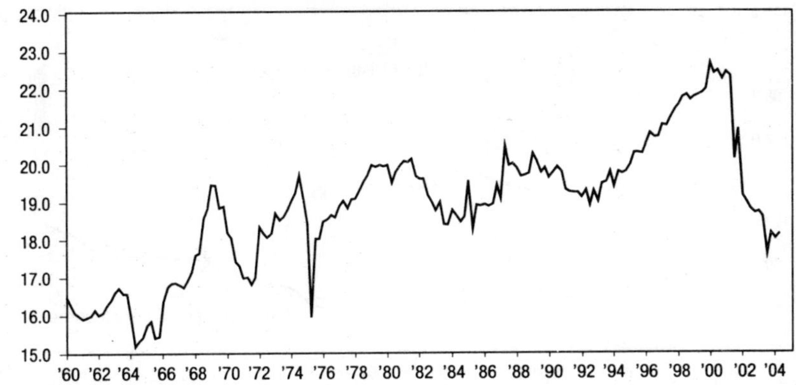

* Percentage of GDP, personal includes current taxes and contributions for government social insurance.

Sources: BEA; CSFB

FIGURE 4.7 **Combined After-Tax Personal Income and Corporate Profit Share of GDP**

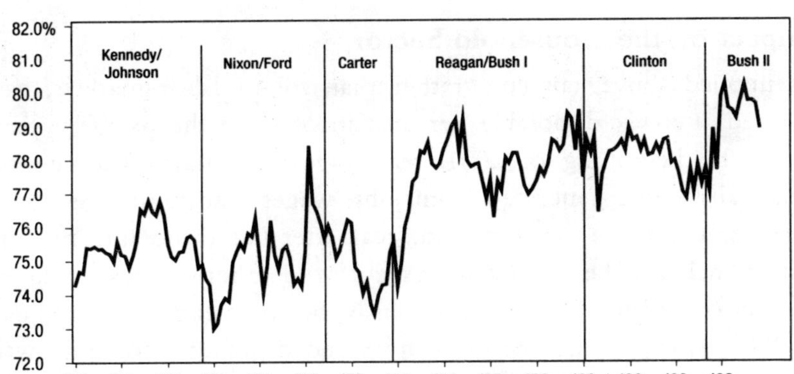

Sources: BEA; CSFB

outright declines in corporate profits during the business downturn. The rest of the decline—particularly during 2003—was due to the tax cuts for households and businesses. A small rise in the share has occurred since then, reflecting stronger tax payments tied to the economic recovery.

Both after-tax personal income and after-tax corporate profits have captured bigger shares of the national income simultaneously (see ***Figure 4.7***). In early 2004, the combined after-tax personal and

FIGURE 4.8 **Tax Cuts to the Rescue**

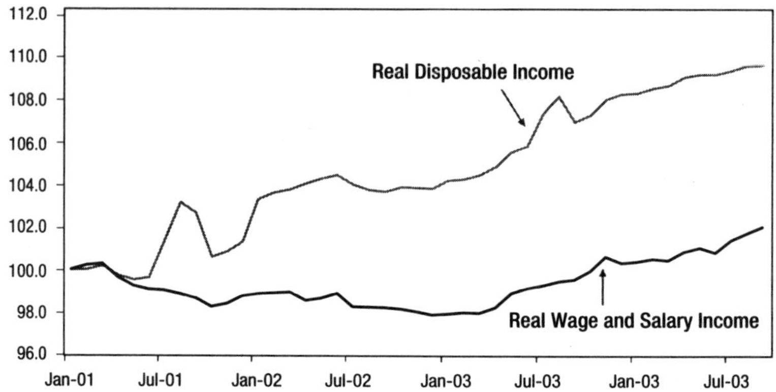

Note: Index levels, Jan 2001=100

Sources: BEA; CSFB

after-tax corporate share of national income was at an all-time high. Both tend to move inversely over time, as profit upswings usually come at least partly at labor's expense, and vice versa.

Impact on the Household Sector

As implied above, tax cuts, rather than robust labor markets, have boosted after-tax disposable personal incomes in the past few years. Since the beginning of 2001, real wage and salary income—the personal income generated from jobs, wages, and hours—is up 1.7 percent. Over the same period, real after-tax disposable income, which includes the impact of tax cuts and other rebates, is up 9.6 percent (see ***Figure 4.8***). Put differently, between early 2001 and early 2004, after-tax income growth exceeded pretax income growth, implying a positive fiscal boost to households' income. Since April 2004, pretax incomes started to grow slightly more rapidly than after-tax incomes, suggesting no new fiscal impulse for households (see ***Figure 4.9*** at right).

The loss of fiscal impulse increasingly puts the responsibility of household cash flow growth—and thereby consumer outlays—on jobs and wages recovery, and on the Federal Reserve's conduct of monetary policy of withdrawing the monetary stimulus only gradually. In terms of growth rates, the CSFB central forecast assumes real disposable personal income increases 3–3.5 percent in 2005. The upside risks to this

FIGURE 4.9 **The Fiscal Impulse on Household Incomes**

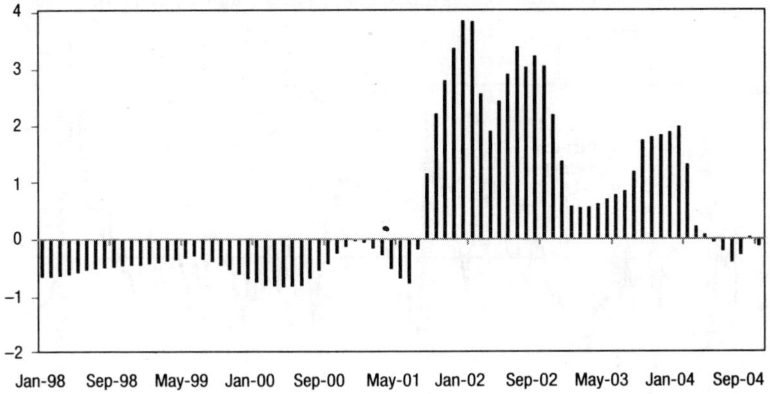

Note: Spread between pretax and after-tax income growth, six-month/six-month annualized.

Sources: BEA; CSFB

projection are a stronger-than-expected acceleration in jobs gains and a steeper-than-expected decline in oil prices.

Impact on the Corporate Sector

The decline in the corporate tax burden in the form of "bonus" depreciation has played an important role, alongside rapid gains in productivity growth, in the recovery of corporate profits. For example, CSFB estimates that the level of after-tax corporate profits was $49 billion higher in the third quarter of 2003 than it would have been with no tax program.

In turn, corporate tax windfalls have played an important role in closing the "financing gap" for the corporate sector. The financing gap is the ratio of capital outlays to internally generated (after-tax) cash flow. The rapid rise in this ratio at the end of the 1990s suggested that capital spending was overextended. As of 2004, the financing gap had dropped *below* its long-run average. After-tax cash flow currently exceeds capital outlays—a positive sign for capital spending.

In sum, via tax cuts, the U.S. Treasury has provided the conduit for the corporate sector to raise after-tax cash flow (see ***Figure 4.10*** on the following page) without issuing debt. In this respect, the commensurate rise of Treasury debt supply, combined with lower (than otherwise) corporate debt supply, has probably been a factor contributing to corporate bond spreads-to-Treasury notes narrowing over the 2003–2004 period.

FIGURE 4.10 **Corporate Financing Gap**
(Nonfinancial corporate sector, ratio of capital outlays/internal funds)

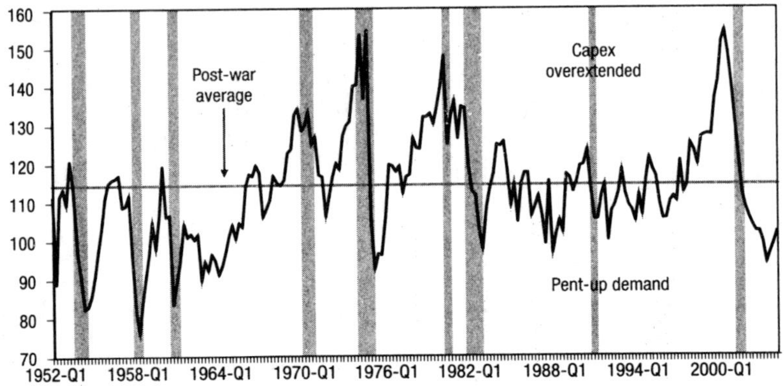

Source: Federal Reserve; CSFB

Impact on the Global Economy

In addition to supporting U.S. growth, the countercyclical U.S. fiscal policy had impressive effects on global growth as well, which underscores the strong policy transmission mechanisms onto the global economy. Macroeconomic models, such as the UK's National Institute Global Econometric Model (NIGEM), give a broad order of magnitude for the effects of the multiyear U.S. stimulus on the global economy. The results must be used with caution, given the well-known shortcomings of such models, but they are interesting. Assuming constant nominal fixed exchange rates, the effects of the tax cut packages on the U.S. economy are estimated to have totaled a cumulative 2.1 percentage points of GDP between early 2002 and year-end 2003, and 2.6 percentage points if the simulation is extended through 2004. In other words, real GDP is estimated to be 2.6 percent higher with the tax cut effects than without them (see ***Figure 4.11***). The model estimates that *global* growth would have been close to 1 percent slower without the U.S. tax cuts. The transmission effects are considerably stronger on Asian growth than on European growth—the United States imports slightly more from China and Japan than from the whole European Union, but exports twice as much to the European Union than to Japan and China. Japan would have registered a cumulative 0.8 percent slower growth by Q4:03 and 1.5 percent by the end of 2004 without the tax cuts, while China would be slower by 0.5 percent and 0.9 percent,

FIGURE 4.11 **Real GDP Growth by Region without Effects of U.S. Tax Cuts**

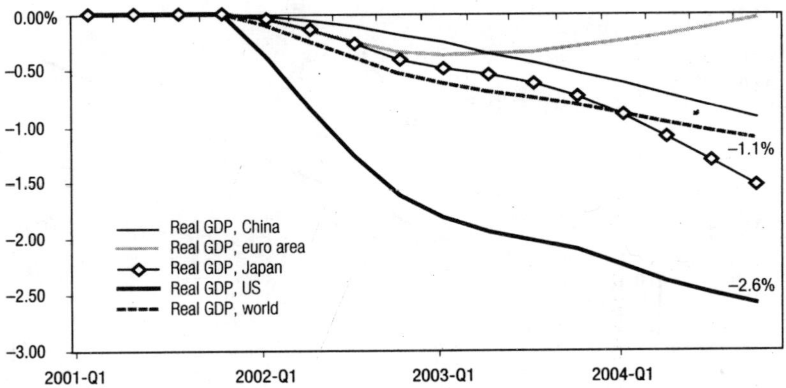

Source: NIGEM; CSFB.

respectively. The impact on the euro area is considerably smaller, at 0.3 percent by Q4:03 and close to no impact by the end of 2004.

The combination of the very high propensity to consume and to import makes the United States the perfect fiscal "borrower of last resort" for global growth (see ***Figure 4.12***). This has yielded a strong appetite for world savings. The IMF estimated that in 2003 the United States absorbed 74.2 percent of global capital—a trend that has continued strongly into 2004—and is likely to continue as long as foreign investors expect that the return on capital is greater in the United States than elsewhere (see ***Figure 4.13***). The largest exporter of capital in 2003 was Japan, with close to 21 percent of the global supply. Germany and France posted a combined total of 13.5 percent.

The flipside of U.S. fiscal largesse has been a more fiscally restrictive rest of the world. Euro area fiscal policy is framed within the Stability Pact's targets of 3 percent deficit-to-GDP and 60 percent debt-to-GDP ratios. Since early 2003 the euro area's fiscal policy has been mildly restrictive even with the negative output gap, but meeting the Stability Pact's targets has proven difficult (see ***Figure 4.14***), and regional differences prevail. For example, France had a slightly countercyclical fiscal policy, while Germany had adopted restrictive measures, but registered a cyclically induced deterioration in the deficit. Italy adopted one-off measures, while Spain maintained its balanced budget. Looking ahead into the 2005–2007 period, European Union member states project to

FIGURE 4.12 **The U.S.'s Growing Appetite for Imports**
(U.S.: Import penetration [ex-Oil])

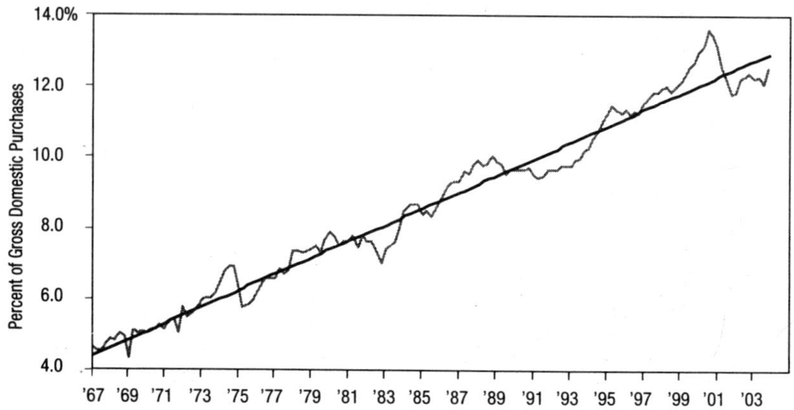

Source: Datastream International Limited; CSFB

FIGURE 4.13 **Strong U.S. Appetite for Foreign Savings**
(U.S.: Bond inflows/12-month moving average)

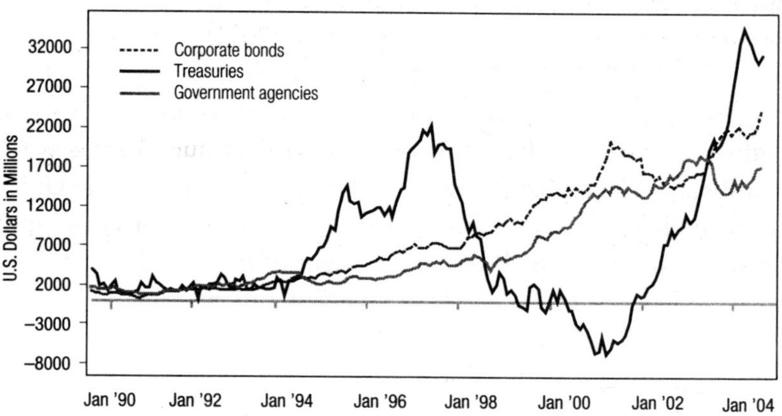

Source: U.S. Treasury; CSFB

bring collectively a deficit reduction of about 0.5 percentage points per year, largely reflecting spending cuts. But this improvement might fall short of bringing the deficit down in the event GDP growth is slower than baseline. That said, structural reforms are being adopted, with the focus on the sustainability of social security systems (pension and health care reforms in particular) across Europe, following ongoing reform measures targeting European labor markets.

Fiscal consolidation is also in place in Asia in 2004. In Japan, fiscal restraint is part of the government's key structural reform program. For 2004, the government will rein in growth in spending, with general expenditure up only 0.1 percent for the year. Public works will fall for the fourth consecutive year, while local government grants will fall 5.2 percent for the year. Tax revenues are expected to remain very sluggish, reflecting tax cuts and the weak economy of the past decade. The level is nearly as low as in FY1986 (pre-bubble era). Therefore, the deficit-to-GDP ratio is likely to remain at 7.2 percent for 2004, despite the restrictive efforts and the relatively firmer economic growth environment.

Fiscal policy will likely be neutral or will tighten across Southeast Asia through 2005. In Malaysia and, to a lesser extent, Indonesia and Singapore, the fiscal deficit as a share of GDP in 2004 will fall enough to generate a slightly restrictive fiscal policy stance. Deficits in these countries should contract at an even faster pace in 2005. Thailand is boosting spending in 2004 to maintain a neutral fiscal impulse, but will probably turn policy restrictive in 2005, as the output gap closes further. The Philippine government has planned a tighter fiscal position as a share of GDP for 2004 and 2005, and India should be able to manage a very gradual improvement in its fiscal deficit, but at a pace that prevents any meaningful negative impulse.

Finally, China's fiscal stance has shifted to a slight tightening from 2003 that will continue through 2005. The strategy is for the government to put fiscal policy in "stand-by" mode while tightening monetary policy. If the economy slows down too fast, fiscal policy would jump to the rescue, but the overall tone should be one of tightening. Hong Kong and Taiwan are likely to be neutral, while South Korea should be more expansive and use fiscal policy to support growth.

The State of Fiscal Policy

With the rest of the world aiming to achieve some degree of fiscal consolidation, it is clear that the longer the U.S. policy assist, the more sustainable the global expansion should become. The risk, however, is for the United States simply to go on spending and borrowing. That is an unstable situation, and the economic consequences would be costly in terms of both the reduction of future national income and future policy flexibility. There have been historical precedents. Between 1980 and 1993,

FIGURE 4.14 **Government Balances in the Euro Area**

% GDP	General govt balance 2002	2003	2004E	2005E
Austria	-0.2	-1.1	-1.1	-1.5
Belgium	0.1	0.4	-0.3	-0.2
Finland	4.3	2.3	2.5	2.5
France	-3.2	-4.1	-3.6	-3.0
Germany	-3.7	-3.8	-3.7	-3.3
Greece	-3.7	-4.6	-5.3	-3.9
Ireland	-0.2	0.1	0.0	0.0
Italy	-2.3	-2.4	-2.9	-3.0
NL	-1.9	-3.2	-3.0	-2.7
Portugal	-2.7	-2.8	-2.9	-3.0
Spain	-0.1	0.4	-0.7	0.0
E12	-2.4	-2.7	-2.8	-2.6
UK	-1.7	-3.3	-2.9	-2.7
US	-3.4	-4.6	-4.9	-4.3
Japan	-7.4	-7.7	-7.2	-5.9

the U.S. government debt held by the public rose as a share of GDP from 26 percent to just below 50 percent. It then declined to 33 percent by FY01, thanks largely to the surpluses achieved in the late 1990s. It took the fiscal tightening of 1990 (G. H.W. Bush) and 1993 (Clinton) and the major stock market rally to achieve these surpluses. In periods of war, this pattern is expected to be broken, at least temporarily, with the risk of a rapid rise in the debt-to-GDP ratio over the medium to long run. As long as interest rates are low, the interest payment on the debt rises only slowly. But medium-term fiscal policy should aim to orchestrate a transition from an extremely lax fiscal policy to at least a stable one, which would target a steady *deficit*-to-GDP ratio. Even then, debt (held by the public) sustainability will not be achieved—that is, a policy that stabilizes or reduces *debt*-to-GDP ratios. If the deficit-to-GDP ratio reaches 3.5 percent by 2005 and stays stable for the remainder of the decade, the debt-to-GDP ratio still would rise from the current 36.1 percent to 46.5 percent by 2012. For the debt-to-GDP ratio to remain stable, say in the 37–38 percent range, the deficit-to-GDP ratio needs to decline to 2–3

"Structural" balance				Chg. in structural bal. pp				Debt
2002	2003	2004E	2005E	2002	2003	2004E	2005E	2003
-0.2	-0.8	-0.8	-1.3	-0.6	-0.6	0.0	-0.5	65
0.0	0.1	-0.5	-0.5	0.5	0.1	-0.6	-0.1	101
3.8	2.3	2.5	2.6	0.6	-1.5	0.3	0.0	45
-3.7	-3.9	-3.6	-3.4	-1.1	-0.2	0.3	0.2	63
-3.4	-2.6	-2.5	-2.0	-0.3	0.8	0.1	0.5	64
-1.5	-1.3	-2.3	-0.9	0.5	0.2	-1.0	1.4	109
-1.9	-0.3	-0.4	-0.3	-0.9	1.6	-0.2	0.1	32
-2.3	-2.1	-1.8	-1.8	0.5	0.2	0.4	0.0	106
-2.1	-1.5	-0.8	-0.2	0.0	0.6	0.7	0.6	55
-2.7	-1.5	-1.4	-1.4	1.1	1.2	0.1	0.0	59
-0.2	0.3	0.0	-0.2	0.4	0.5	-0.3	-0.2	51
-2.4	-2.4	-2.4	-2.1	0.0	0.0	0.0	0.2	70
-1.4	-2.7	-2.6	-2.5	-1.8	-1.3	0.2	0.0	40
-2.7	-3.2	-4.0	-3.7	-1.6	-0.5	-0.8	0.3	63
-6.3	-7.5	-7.0	-6.0	-6.3	-0.5	0.5	1.0	155

Sources: CSFB

percent, implying a sharp reduction in the budget deficit on the order of $100 billion over a period of a year to eighteen months.

Deficit reduction of that magnitude entails hard policy choices and the risk of an unwanted economic slowdown. Foreign capital flows and the ability of the United States to attract foreign capital to finance the fiscal imbalance help postpone these choices, as the shortfall in domestic savings is being partially made up by foreign savings.

How stable are these flows? Foreign flows tend to be of two types. Foreign direct investments are the most stable form of financing generally geared at longer-term investments in productive assets; and portfolio flows into bonds and stocks tend to a more volatile form of financing the lack of savings in a country. Portfolio flows into bonds now dominate foreign capital inflows into the United States and reflect two major trends. The first trend is flows from Asia that are tied to Asia's export-led economic growth model. They mirror the trade flows, and the growth model appears well in place, ensuring capital flows into the United States, even with a currency regime change in China. Foreign exchange intervention adds to the dynamic, as intervention to prevent the home currency from appreciating against the dollar (in order to ensure ongoing exports to the United States) means buying dollars and reinvesting them in U.S. Treasuries. For example, Japan's massive foreign exchange intervention in 2003 added substantially to the already large Japanese holdings of Treasuries.

The second trend is related largely to flows from Europe seeking yield and duration in a more liquid and diversified U.S. market. They reflect changes in pension rules in Europe (which caused firms there to scale down equity investments and match liabilities more with fixed-income investments) and demand from insurance companies. As long as foreign investors expect the return on capital to be greater in the United States than elsewhere, capital inflows to the United States will continue, and that, in turn, provides the United States with more policy flexibility than otherwise would be the case, or less of an incentive to increase domestic savings. But the slide in the dollar over the past year is a reminder that foreign investors increasingly are demanding a price concession for their purchases of U.S. assets.

Conclusion

There is no question that the U.S. fiscal position has swung dramatically from 2001 to 2004 and that the legislative changes over this period are by far the more important factor having contributed to the shift from budget surpluses to deficits. The ten-year cost of the three tax-cut packages enacted over these three years is considerable, estimated to total $1.74 trillion. Importantly, though, the role of U.S. fiscal policy over the past four years has been instrumental in buffering the cyclical downturn not just in the United States but also in the global economy. With globalization and thus rising economic integration, policy changes in the United States have a large impact on economic activity in other countries, and the role of the United States as locomotive to global growth remains firmly in place. As such, the combination of the very high propensity to consume and to import makes the United States the perfect fiscal "borrower of last resort" for global growth. This has yielded a strong appetite for world savings. But with global growth close to a thirty-year high, fiscal policy should aim to orchestrate a transition from an extremely lax fiscal policy to at least a stable one.

5

Globalization and Trade—The Role of Proximity, Borders, and Culture

TIM O'NEILL

TIM O'NEILL speaks quietly. When the topic is the tensions of border and culture, economists lean forward and take in every word. As the former chief economist at BMO Financial Group (formerly Bank of Montreal), O'Neill brings a Canadian perspective to a too-often American-centric field of analysis. His optimistic economics is based on behind-the-scenes quantitative research, analysis that can surprise the most sophisticated reader. Here, O'Neill destroys pop-globalization myths and rebuilds a foundation of interdependent trade realities—realities of nations and people grounded in a world's timeless need to trade and, perhaps, trade freely.

There is a perception among many—both supporters and opponents—that globalization is so pervasive that individual nation-states and their citizens are relatively powerless to withstand its onslaught. Increasing interdependence engendered by ever-rising flows of trade, capital, and people renders the world a global village in which independent economic and social policy is more and more constrained by "outside" influences. The impact on domestic inflation, employment, fiscal measures, social policy, and the like is such that public policy-makers have little room to maneuver. Firms and individuals as well are increasingly constrained in their decisions and behaviors by competitive pressures from global economic integration.

However widespread this view is, the actual facts about globalization clearly belie it. First, trade patterns have, over the past twenty-five years, become more regional and less global rather than the reverse. Hence, proximity still matters even in a world where transportation costs have dramatically declined. Second, national borders remain important even as policy barriers to movement across borders have diminished. Finally, the reason that distance and borders still matter probably has more to do with non-economic, nonpolicy or "natural" barriers—linked to norms, networks, and information asymmetries—than to factors that can be readily influenced by public policy.

Globalization and Trade—Preliminary Issues

The term "globalization" is so weighed down with misconceptions and misrepresentation that a clear definition is needed. Globalization, in the context of this chapter, describes the increasing economic integration or linkages among countries and groupings of countries. The linkages comprise flows of trade, capital (both portfolio and direct), labor, and technology. While there may be institutional and social manifestations of integration—such as increasing cross-border similarities in laws and regulations governing commercial activity, and in social policies—these are likely outcomes rather than constituent elements of integration. In fact, as shall be argued later, the extent to which such convergence has occurred is quite limited.

The most widely used indicator of cross-border economic integration focuses on the primary avenue of access to foreign markets—that is, international trade in goods and services. Internationally, trade has tended to grow faster than domestic production, meaning that countries are increasingly reliant on each other to produce goods and services consumed by their citizens.

Countries are also linked economically by capital flows. Portfolio investment involves international transactions in equities and fixed-income securities. When portfolio investment in a particular firm crosses the management control threshold, it is classified as direct investment. Direct investment can also involve outright expenditures on plant and equipment. Foreigners, thereby, add to the productive capacity of a domestic economy.

Finally, integration can occur by way of people moving (temporarily or permanently) from one country to another to work. Generally speaking, labor market flows are very small relative to trade and capital movements, owing mainly to the still extensive restrictions on inter-country migration. Even where such impediments are reduced—for example, within the European Union—language barriers can be a significant impediment to international labor mobility.

Regionalization versus Globalization?

International economic integration has tended recently to be increasingly a regional rather than a global phenomenon. That is, the linkages among the world's economies have tended to be stronger within defined geographic groupings or blocs than among these regions. The three dominant blocs—North America (NAFTA), the European Union (EU), and the broad Asian region (Japan and "Developing Asia")—together account for about 80 percent of global GDP. Within those blocs, intra-regional trade in goods now accounts for 50 percent or more of exports and, for two of the regions, the ratio has increased significantly over the past twenty years. Intra-NAFTA exports were 56 percent of total NAFTA exports in 2000, up from 34 percent in the early 1980s. Intra-Asian exports were 50 percent of total exports in 2000, well above the 35 percent range exhibited during the early 1980s. Intra-EU exports were 63 percent of total exports, having sustained this level during most of the past twenty years.

It is perhaps not surprising that there should be a trend toward "regionalization" of trade since, along with multilateral trade liberalization, there have also been regional trade agreements such as the EU and NAFTA. However, it is worth noting that Asia, with very limited regional liberalization, has displayed the same pattern of regionalization. That suggests that regional trade agreements probably reinforce a tendency established by geographic proximity—in other words, they "ratify" trade patterns that already exist.

There is another set of data that supports the "regionalization" thesis and captures the joint impact of trade and foreign direct investment activities that firms use to access foreign markets. Multinational enterprises (MNEs) are involved in both exporting domestic production and establishing foreign affiliates to produce and sell in external markets. If there were evidence of globalization, it is in their activities where it should be most evident. However, the geographic distribution of sales of the world's 100 largest MNEs also shows a clear intra-regional pattern. Of the eighty-two MNEs for which complete regional sales data are available, sixty-eight have at least 50 percent of their sales within their home region. Only three MNEs are truly global (defined as at least 20 percent of sales in each of the three regions).

This is a powerful result, given that these firms would tend to command a significant share of global trade, both inter-regionally and intra-regionally. Were the list expanded to include firms further down the list (in the top 500 or 1,000), the intra-regional share of sales would very likely increase as smaller firms would tend to be even more oriented to their home regions.

That regionalization tendencies are influenced by geographic proximity, despite a persistent long-term decline in transportation costs, is borne out by the research done on trade using so-called "gravity models." Such work consistently demonstrates that trade between any two geographic areas is directly proportional to their relative sizes and inversely proportional to the distance between them after adjusting for relevant transport costs. This begs the question of why proximity should matter.

A key advantage of proximity is similarity of and/or knowledge about customer behavior—such as tastes and preferences, along with sensitivity to price and quality changes. This is not as pertinent an issue for products that are standardized or commoditized (such as producer inputs and basic consumer goods). However, where product characteristics (including how and where delivered) are important to buyers, knowledge of customer behavior can be advantageous to a seller. Contiguous countries are more likely to have broadly similar patterns, and producers/sellers in those countries are likely to have a more extensive knowledge of those behaviors. This makes customer service requirements easier and less costly to provide.

Institutional structures (such as regulations, laws, and policies) and practices (such as labor market and business "culture") may also affect the behavior of customers along with the costs of operating. Again, jurisdictions that are close to each other are more likely to have comparable structures and practices, making it easier for firms in one jurisdiction to sell into another.

However, as strong a pull to integration as there is among spatially proximate firms and individuals in different countries, there is evidence that there is also a strong pull from within each country. That is, the factors that support regionalization in integration are even more relevant to "localization."

National (Border) Effects—Globalization versus "Localization"‡

Contrary to views of both proponents and opponents of international economic integration—global and/or regional—what is (arguably) surprising is that such cross-border integration is not considerably more extensive. That is, there appear to be significant, nonpolicy barriers to integration that can be captured under the rubric of "national or border effects."

A national or border effect on trade exists if flows of goods and services are larger within countries than between them after adjusting for distances between subnational regions and their relative sizes. A study of Canada-U.S. trade flows prior to the Canada-U.S. Free Trade Agreement in 1988 found, for example, that merchandise trade flows between Ontario and British Columbia were twice as large as between Ontario and California.[1] Since California's economy is ten times the size of British Columbia's and is equidistant from Ontario, gravity models would have predicted that Ontario-California trade would be ten times larger than Ontario-British Columbia flows. Hence, the Ontario-British Columbia trade pattern was twenty times larger (a border effect of 20) than would be expected. For Canada-U.S. trade as whole, the border effect in 1988 was 17. An update of the analysis indicates that, in the aftermath of the Canada-U.S. Trade Agreement (CUSTA), the border effect has declined significantly but remains a still-sizeable 12.

Similar empirical work on other major trading areas has been carried out, albeit using less complete intra-country data and utilizing input-output data. These studies indicate that there are material border effects both for the industrialized countries and for developing countries. In fact, this "natural" barrier appears to be larger for the latter group. "These effects are still large and significant even for trade between pairs of countries that have long been members of the European Union," writes John F. Helliwell in *Globalization and Well-Being* (UBC Press, 2002).

The existence of border effects can also be inferred from empirical work which shows that despite steadily rising international trade flows, the "law of one price" does not apply in international markets. For

‡ This section borrows extensively from Helliwell (2002).

example, a direct test of cross-border price arbitrage between Canada and the United States shows that product price alignment occurs more extensively within countries than between them.[2] More generally, studies of purchasing power parity indicate that after adjusting for the impact of exchange rate changes on domestic prices, there is no tendency for international transactions to cause convergence to a single price for traded goods and services.

Not incidentally, there is also strong evidence of border effects for capital markets. More than twenty years ago, Feldstein and Horioka—utilizing inter-country patterns of domestic savings and investment rates—concluded that capital markets were more national than international.[3] Subsequent research along the same lines has tended to confirm the initial results. Moreover, empirical analysis of portfolio capital movements has persistently demonstrated the existence of a strong home-country bias among investors.[4]

What Accounts for Localization?

The evidence, in sum, suggests that distance matters, irrespective of transportation costs, so that spatial proximity of countries influences trade and sales patterns over and above regional trade liberalization. It also seems clear that borders matter, irrespective of distance. Is it possible to account for both regionalization and localization?

Helliwell has argued that "large international differences in norms and networks" tend to effectively separate countries in their respective economic activities.[5] He argues that local (within country) connections and shared norms (including formal institutions and legal systems) bind individuals and firms together more closely than is commonly supposed, even if they are separated by significant distances. Internal migration and travel tend to be more extensive than inter-country movements, hence adding to the degree of intra-country familiarity and comfort.

In *The Well-Being of Nations: The Role of Human & Social Capital* (OECD, 2001), the OECD uses the term "social capital" to refer to the "networks together with shared norms, values and understanding that facilitate cooperation within or among groups." This infrastructure of individual and institutional interconnections is clearly different from the variables that economists typically use to analyze an economy's performance. The persistence of border effects suggests, however, that they are relevant. Although the term has a number of meanings and connotations, it is appropriate to refer to these variables collectively as "cultural" factors to distinguish them from more conventional economic and political factors.

Within a country, the informal connections among individuals, groups, firms and institutions generate understanding and trust that increase the willingness to engage in a wide range of transactions including but not limited to commercial ones. Individuals and groups within a country may have widely differing political and social views but are likely, nonetheless, to have a set of underlying values or norms and customs that are broadly shared and accepted. As well, over time, formal institutions develop and evolve that intersect with and reinforce the informal structures. National institutions, policies, and legal frameworks are among the components of these formal structures.

While they are posited to explain border effects, Helliwell's arguments regarding networks and norms can be generalized to capture both localization and regionalization patterns. It is reasonable to suppose that while norms and networks are stronger within countries than between them, the same would be true for intra-regional versus inter-regional patterns as well. That is, distance and culture are likely to interact not only within countries but within regions as well.

Customs and norms are likely to be more similar among contiguous countries than for countries far apart geographically. Institutional structures—legal, regulatory, and policy-making—and practices (such as labor market behavior and business "culture") tend to be more comparable. Spatially proximate jurisdictions will have more shared history and experiences—even if the sharing hasn't always been amicable. Smaller countries may also be more amenable to—or, at least, more influenced by—larger neighbors' cultural norms.

Networks are also more readily formed across the borders of neighboring countries than over an ocean or widely separated land masses. In North America, Europe, and Asia, cross-border regional political and economic associations have emerged based on perceived common problems and interests.

To Helliwell's networks and norms I would add a third factor that affects regionalization and globalization. Information asymmetries are likely to increase over distance and across borders. As noted, firms' knowledge about customer behavior—such as tastes and preferences or sensitivity to price and quality changes—can be a critical factor in attracting and retaining buyers. Customer behavior patterns are likely to be more similar, and knowledge about them more extensive, within countries and regions than between them. From a customer perspective, brand recognition and trust will display comparable intra-country and intra-regional patterns.

Helliwell alludes to this when he suggests that, among the factors preventing international trade patterns from reflecting comparative advantage are "costly information, diverging knowledge and tastes and transactions costs that grow substantially with distance."[6] In a similar vein, Maurice Obstfeld, assessing the home-country bias of equity investors, argues that "unfamiliarity with foreign products, firms, business practices, accounting standards, political trends, and regulatory environments surely plays some role."[7]

The work to date on border effects suggests that globalization, as it has been conventionally perceived, is far less extensive than detractors and supporters alike have supposed. Rather, regionalization and localization remain pervasive limitations on international economic integration. Nonetheless, the pre- and post-CUSTA evidence for Canada does imply that border effects are subject to erosion in the face of trade liberalization.

Policy Implications of Regionalization and Localization

International economic integration makes it easier for firms to shift or expand operations in countries where market and/or cost conditions are more attractive. However, the role of both distance and border effects suggest there are significant constraints on the likely magnitude of such shifts. For example, property rights and enforceability of contracts are key institutional elements of an attractive investment environment.

Even if firms are outsourcing services to operations outside their home country (that is, importing services from abroad), the uncertainty about foreign norms, networks, infrastructure, and local skills, along with the challenges of managing across distances and languages, will limit the extent to which such outsourcing will grow. As has almost invariably been the case in the past—recall the political hype over the "giant sucking sound" of jobs moving to Mexico from the United States—the magnitude of any such job shifting will tend to be radically overestimated.

The much broader issue about domestic policy tends to be about the impact of globalization and international trade competitiveness on a country's capacity to maintain sovereignty in the face of global economic pressures. This tends to be much more of an issue in small open economies, such as Canada, than in large, more closed economies, like the United States. The following discussion uses the Canadian case as an example of the perceptions and reality for small countries.

Domestic Macroeconomic Policy

With respect to macroeconomic policy, economic theory does not suggest that rising international economic integration will limit the effectiveness of monetary and fiscal policy except in the case of currency union. The Canadian experience has, in fact, been one in which, over protracted periods, monetary and fiscal policy have operated quite differently than in the United States. This is despite the fact that the Canada-U.S. economic integration has increased dramatically over the past thirty years. Canada currently exports 40 percent of its output, and 85 percent of that goes to the United States, so that about one-third of its production is directly linked to its major trading partner. In addition, the United States is the biggest source of direct investment in Canada, and there is considerable cross-border integration within a number of manufacturing industries (such as machinery, transportation equipment, computer and electronic products, and electrical equipment).

Differences in monetary policy, measured either by broad monetary conditions for the two countries or (more simply) interest rate spreads, show quite substantial differences in monetary policy stances pursued by Canada and the United States over the past fifteen years. For example, in the late 1980s and early 1990s, Canada maintained a much tighter monetary stance and a much looser fiscal posture than did the United States. The result was interest rates substantially higher than corresponding U.S. rates and a much deeper recession in 1990–1991, followed by a much slower recovery. In the latter half of the 1990s, monetary policy in Canada was initially far more stimulative than in the United States, and interest rates fell well below corresponding U.S. rates, but gradually moved back in line with U.S. rates as both economies operated at full capacity. Since the downturn in 2001, the relatively stronger Canadian performance has meant a higher rate environment in Canada. In early 2004, a 100 basis-point difference in rates existed, with Canada above the United States.

On the fiscal front, the past decade provides a clear example of the ability of the Canadian federal government to pursue a fiscal course of action quite distinct from that of the United States. In particular, the spiraling debt burdens of Canadian governments in the late 1980s and early 1990s forced Canada into a far more restrictive fiscal stance than the United States through the latter half of the 1990s. Since 2001, the Canadian fiscal stance has been essentially neutral to mildly restrictive, while U.S. fiscal policy is deliberately expansionary in an attempt to stimulate the economy.

One area where increased integration may arguably have reduced flexibility is in tax policy. Clearly, with the greater prospective mobility of capital, domestic corporate and capital gains tax policies must be sensitive to the impact that higher domestic tax rates will have on the potential outflow of capital. Indeed, in recognition of this, the Canadian government lowered the capital gains tax rate to essentially U.S. rates and planned to lower corporate tax rates from about the U.S. level to slightly below by 2004. With respect to personal tax rates, the fact that labor is far less mobile between countries than capital has allowed substantial differences to continue. This is unlikely to change in the near term.

Beyond the Canada-U.S. comparison, what do we observe about fiscal policy, in general, and taxation in particular? Without attempting a comprehensive examination of the evidence, certain obvious patterns indicate that individual countries have considerable latitude to pursue quite different policies. Looking at government spending, countries vary considerably in the share of total economic activity that government expenditures constitute. Within the OECD, government outlays, as a share of GDP, range from a high of 53 percent in Sweden to a low of 23 percent in South Korea. The United States is third from the bottom at 30 percent.

With respect to tax burden, both the overall level and its composition—the split between taxes on labor and on capital—are relevant. The dispersion in tax burden levels is consistent with that for government spending. However, there is evidence of international competition with respect to taxes on capital—both tax rates in general and the use of preferential taxes for attracting foreign direct investment. In a 1998 article in the *Journal of Economic Perspectives,* Dani Rodrik concludes that openness has led to a shifting of some of the tax burden from capital to labor as competitive pressure leads to capital-tax leveling.[8] As mentioned above, this is a logical outcome of the greater mobility of capital than labor. However, since increased capital investment is widely agreed to have a long-term impact on productivity, growth, and living standards, a policy of reducing taxes on capital have merit in their own right, irrespective of the pressures from globalization.

Domestic Social Policy

With respect to the conduct of social policy in the context of globalization, several general observations will have to suffice.

Canada, like all other countries, has long established the main elements of its labor market policies, and there is no evidence that international integration has yet had any effect on such policies. In fact, countries vary widely in the extent to which they attempt to inhibit firms from laying off workers. Compensation programs (such as benefit levels, eligibility rules, training programs) for the unemployed also differ across borders. Some governments even incorporate regional and sectoral variations in their unemployment policies. And there are significant restrictions on the legislation governing the role of unions, including those for certification processes and for membership requirements. What is perhaps most striking is how little obvious influence integration has had on labor policies. Witness, in particular, the differences among members of the European Union, which have become increasingly interdependent over the past half century. Although one of the hopes (of economists, at least) was that the introduction of the euro would induce a reduction in labor market rigidities in Europe through restructured rules and regulations, to date there is little evidence of such movement.

In education and in health—the two social policy areas toward which national (and subnational) governments devote significant effort and resources—there is considerable agreement on the desirability of material public sector involvement. (The United States is something of an exception with respect to health care.) Yet, the dispersion among countries in the share of government expenditures or, more broadly, GDP devoted to these is quite wide. Even leaving the United States out of the picture, there are considerable inter-country variations in the public/private split on funding for health and postsecondary education. Finally, health care in particular, but education as well, are subject to considerable differences in the structures for administration and management and in the rules and regulations under which they operate.

The key point is that countries have not in any discernible way become more constrained in the manner or the extent to which they conduct economic and social policies. This is (perhaps) predictably the case for monetary policy. In a world of flexible exchange rates, central banks are free to set interest rates to achieve domestic targets. However, this flexibility would not be so readily expected in other economic policy areas—such as fiscal (especially tax) and labor market policies—let alone for social policies.

To considerable extent, this is explained by the fact that features of globalization are not really as global as is often argued or imagined. At the very least, trade patterns are regional rather than truly international, indicating that geographic proximity matters. It also appears to be the case that economic activity is still influenced by the existence of borders even as cross-border policy barriers are removed. In other words, "culture"—values, formal and informal networks, social policy structures—has a considerable impact on the form and extent of cross-border transactions.

REFERENCES

BMO Financial Group Economics Department. 2002. *North American Economic Integration*. Toronto: Bank of Montreal.

Helliwell, John F. 1998. *How Much Do National Borders Matter?* Washington, DC: Brookings Institution Press.

Ostry, Sylvia. 1998. "Globalization and the Nation State." *Room to Manoeuvre? Globalization and Policy Convergence*. Montreal: McGill-Queen's University Press, pp. 57–65.

CHAPTER NOTES

1. McCallum, John. "National Borders Matter: Canada-US Regional Trade Patterns." *American Economic Review,* 1995, vol. 85, pp. 615–23.

2. Engel, Charles and J. H. Rogers. "How Wide Is the Border?" *American Economics Review,* 1996, vol. 86, December, pp. 1112–25.

3. Feldstein, Martin S. and Charles Horioka. "Domestic Savings and International Capital Flows."*Economic Journal,* June 1980, vol. 90, pp. 314–29.

4. Lewis, Karen. "Puzzles in International Finance," in Gene Grossman and Kenneth Rogoff, eds. *Handbook of International Economics,* vol. 3. Amsterdam: North-Holland, 1995, pp. 1913–71.

5. Helliwell, John F. *Globalization and Well-Being*. Vancouver, BC: UBC Press, 2002, p. 55.
6. Ibid., p. 25.
7. Obstfeld, Maurice. "The Global Capital Market: Benefactor or Menace?" *Journal of Economic Perspectives*. 1998, vol. 12, pp. 9–30.
8. Rodrik, Dani. "Symposium on Globalization in Perspective: An Introduction." *Journal of Economic Perspectives*. 1998, vol. 12, pp. 3–8.

6

The Rise and Fall of the Dollar

MICHAEL R. ROSENBERG

MICHAEL ROSENBERG has reached the pinnacle of Internet acclaim. His 1995 book, Currency Forecasting: A Guide to Fundamental and Technical Models of Exchange Rate Determination, *is a classic. If you can find a used copy online, you may have to pay $175 or more for it. Rosenberg was global head of foreign exchange research at Deutsche Bank. He's gone off to the land of hedge funds but leaves us this chapter as a parting gift. If dollars, yen, and euros are the litmus paper of the world economy, then the chemical experiment itself is trade. Here is Michael Rosenberg on foreign exchange, trade, and an asymmetric experiment-gone-wrong—the U.S. current account deficit.*

It has been said that the economic and financial events of today never exactly repeat those of yesteryear, but they do often rhyme. This is especially true of the movements in the U.S. dollar. Historically, the dollar has exhibited a tendency to both rise and fall over long-term cycles. In a typical dollar cycle, the length and magnitude of the dollar's up and down swings have tended to be large, and in each case, the dollar has shown a tendency to overshoot its purchasing power parity (PPP) level on both sides by rather wide margins. Since the beginning of floating exchange rates in 1973, the dollar has undergone three major downcycles—in the late 1970s, the second half of the 1980s, and the latest downcycle, which began, for the most part, in early 2002 and is still running today (see ***Figure 6.1*** on the following page).

FIGURE 6.1 **Long-Term Cycle In the Deutschemark/U.S. Dollar Exchange Rate, 1972–2004**

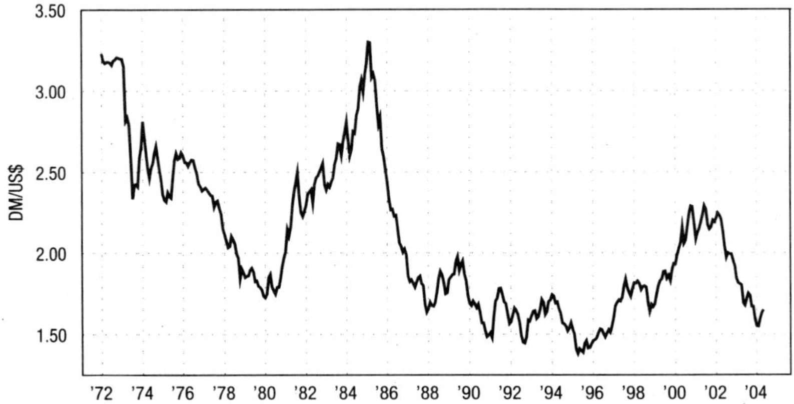

Source: Datastream

FIGURE 6.2 **U.S. Merchandise Trade Balance, 1976–1980**

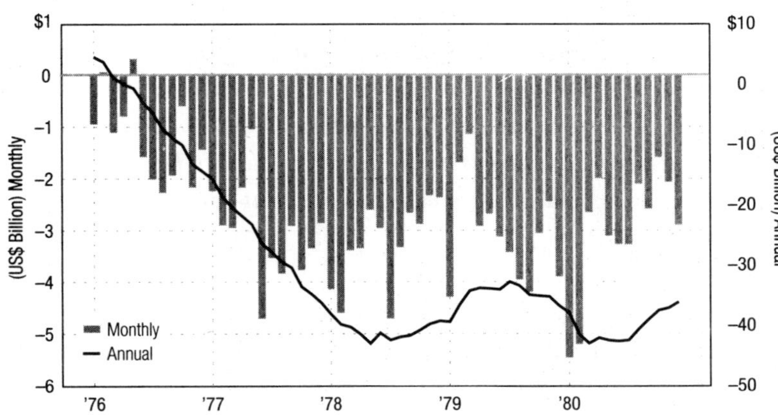

Source: Datastream

Those downcycles share a number of common features. In each case, the dollar downcycle was driven in large part by concerns over outsized U.S. trade imbalances. As shown in ***Figures 6.2*** to ***6.4***, the U.S. trade balance deteriorated dramatically in each episode, which raised concerns among policymakers and market participants alike that the trend in those trade imbalances was not sustainable.

Relatively easy U.S. monetary policies also played a key role in those dollar downcycles. As shown in ***Figure 6.5,*** U.S. real interest rates

FIGURE 6.3 **U.S. Merchandise Trade Balance, 1981–1987**

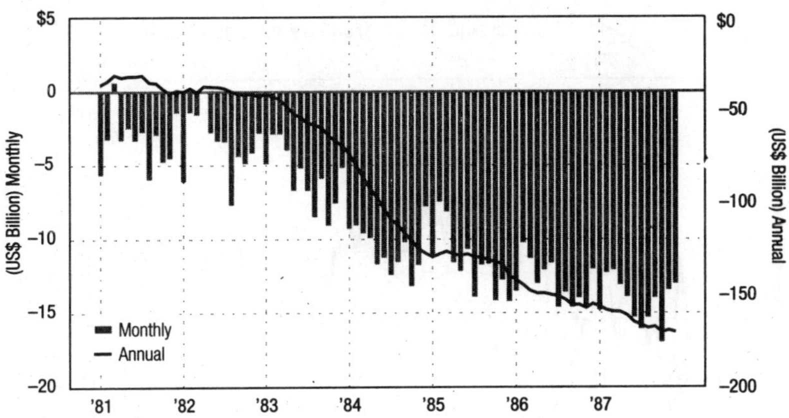

Source: Datastream

FIGURE 6.4 **U.S. Merchandise Trade Balance, 1976–1980**

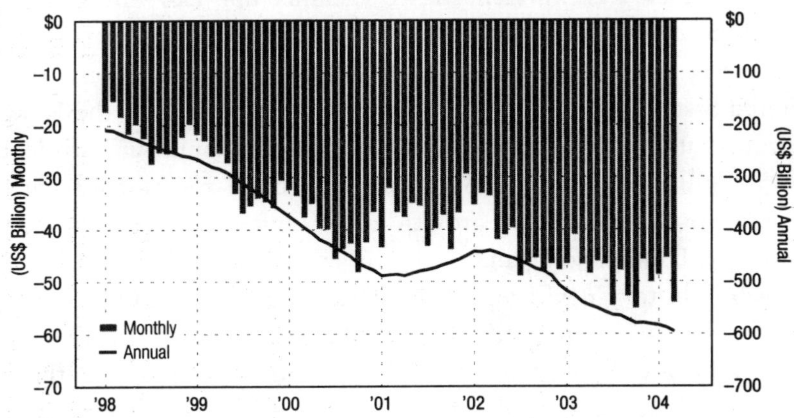

Source: Datastream

were negative during the dollar downcycles of the 1970s and today, and moved down sharply—although they stayed in positive territory—during the dollar downcycle of the second half of the 1980s. Low or negative real interest rates coupled with a massive trade imbalance can be a lethal combination for any currency, and this has been especially true for the dollar over the past thirty years.

Dollar upcycles, like their downcycle counterparts, also share a number of common features. Since the beginning of floating exchange

FIGURE 6.5 **U.S. Real Short- and Long-Term Interest Rates, 1974–2004**
(3-Mo. Euro-$ and 10-Yr. Treasury yield less CPI)

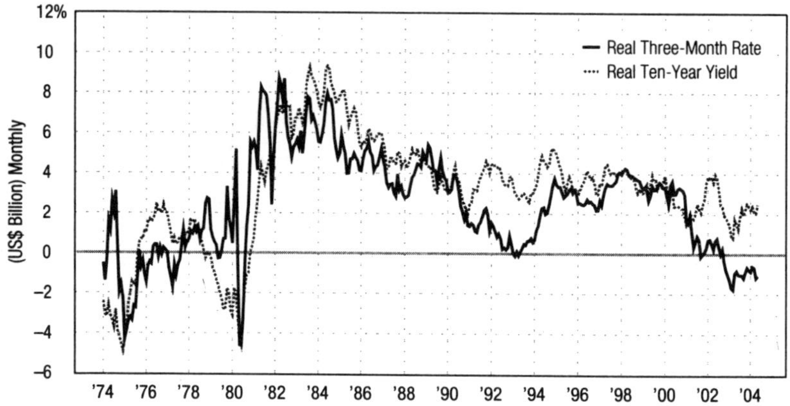

Source: Datastream

rates, the dollar has experienced two sustained upcycles—in 1981–1985 and 1995–2000. Those two upcycles were driven in large part by high and rising real interest rates and sustained economic booms in the United States. The 1981–1985 boom was spearheaded by the Reagan administration's tax cuts and the U.S. defense build-up, while the 1995–2000 boom was driven by "new economy" forces that triggered a large rise in productivity growth in the United States relative to the European Union and Japan.

While history never repeats itself exactly, it is striking how similar the patterns of the dollar's up-and-down cycles of the 1980s and 1995–2004 are. As shown in ***Figure 6.6,*** the dollar's upcycle in the second half of the 1990s tracks the dollar's upcycle in the first half of the 1980s quite closely. Likewise, the dollar's downcycle from early 2002 to the present appears eerily similar to the dollar's dramatic downswing of 1985–1987. A casual observation of Figure 6.6 would indicate that if the dollar's current downcycle conforms closely to the 1980s downcycle, then it probably still has further room to run before it's all over.

One of the purposes of this chapter is to assess just how far and for how long the present dollar downcycle will run. Given the massive size of the U.S. current-account deficit today and the strong likelihood that the U.S. deficit will continue to widen on a trend basis, a considerable further decline in the dollar's value will be needed to promote an

FIGURE 6.6 **A Comparison of the 1980–1987 and 1995–2004 Dollar Cycles**
(Rebased U.S. dollar real broad currency index)

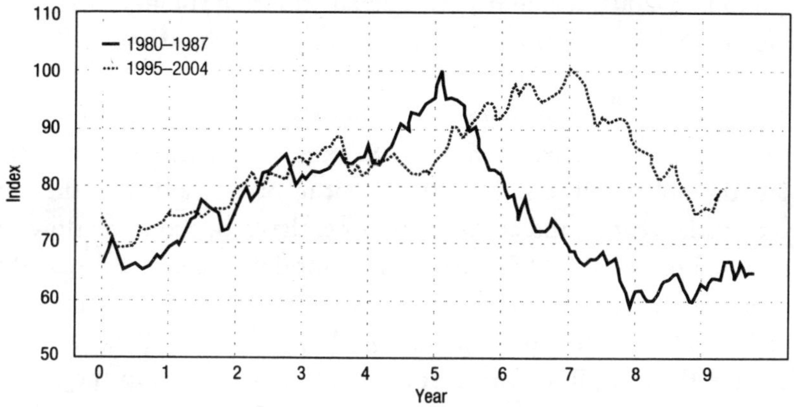

Source: Datastream

orderly adjustment of the U.S. current-account imbalance to a more sustainable level.

This chapter will analyze the fundamental and technical forces that contributed to the dollar's rise in the second half of the 1990s and describe how "new economy" forces contributed to a U.S. investment boom, a surge in U.S. productivity growth, and a significant rise in capital inflows to the United States. Although the U.S. current-account deficit began to widen substantially in response to the sustained rise in U.S. economic activity in the second half of the 1990s, the pace at which private capital was flowing into the United States at that time significantly exceeded the pace at which the U.S. trade balance was deteriorating. This resulted in a net increase in the demand for dollars, which helped propel the dollar sharply higher over the 1995–2000 period.

The chapter then moves on to discuss the fundamental forces that have contributed to the dollar's dramatic decline since early 2002, including:

1. the rise in the U.S. current-account deficit to what is generally believed to be an unsustainably high level;

2. a trend decline in U.S. interest rates to rather low levels, which has made it more difficult for the U.S. to finance its massive current-account imbalance;

3. the need to correct the dollar's overvaluation (the dollar was 25 percent to 30 percent above fair value at its peak by 2000, according to PPP estimates); and

4. a slowdown in net private capital inflows to the United States that began in mid-2001, which necessitated an offsetting rise in official flows to help finance the large and rising U.S. current-account shortfall.

Using a conventional econometric model of the U.S. current account to forecast the future trend in the U.S. external balance, we show that unless the dollar falls another 20 percent to 30 percent over the next two to three years—and at the same time, the U.S. economy slows on a sustained basis—the U.S. current-account deficit appears destined to continue widening to new record levels. Some will certainly argue that a further dollar decline of 20 percent to 30 percent would be excessive. We make the case, however, that a downside overshoot in the dollar's value will not only be necessary to correct the U.S. external imbalance this time around, but in fact exchange-rate overshooting should be viewed as a normal occurrence in a floating exchange-rate system and therefore should be expected and not resisted.

In a world where the pass-through effect of exchange-rate changes on traded-goods prices is limited and where the price elasticities of demand for traded goods are generally low, modest changes in exchange rates are unlikely to be successful in bringing about significant changes in trade imbalances. Rather, it appears that only very large changes in exchange rates are capable of bringing about the changes in trade volumes necessary to correct large external imbalances. Hence, policymakers may need to accept the fact that exchange rates might have to significantly overshoot their long-run equilibrium values from time to time.

Therefore, attempts to manage exchange rates should be avoided because such actions might in the end cause more harm than good, particularly if a dollar-stabilization effort acted to prevent the U.S. trade imbalance from adjusting to a sustainable level. Given that the United States is one of the principal growth engines in today's world economy, it is in no one's interest to have an outsized U.S. external deficit act as a serious constraint on U.S. growth prospects in the future.

The U.S. "New Economy" Boom and the Rise of the Dollar, 1995–2000

Technological revolutions are often associated with financial booms and busts. Major technological revolutions such as the railroad in the nineteenth century and electricity in the early twentieth century gave rise to dramatic improvements in productivity that proved to be sustainable over time. Yet, in both of those instances, there were also associated financial booms and busts. The information technology (IT) revolution of the late twentieth/early twenty-first century has followed a similar script. It led to broad-based gains in U.S. productivity growth that continue to this day, but it also gave rise to a surge in equity share prices in the second half of the 1990s that far exceeded any semblance of fair value. Eventually, U.S. share prices tumbled, restoring equity values to their fair value level.

Equity prices were not the only asset price that got caught up in the IT-induced financial bubble. The U.S. dollar's value also was pushed sharply higher in the second half of the 1990s, and in particular over the 1999–2000 period. Like the U.S. equity market, valuation readings were pushed to rather extreme levels. Our PPP estimates suggest that the dollar reached PPP overvaluation readings of 25 percent to 30 percent at its peak in 2000.

To fully appreciate the role that the "new economy" IT revolution played in the dramatic rise of the dollar during 1995–2000, it is instructive to break down the dollar's rise over that period into two phases: the first one, which ran from the spring of 1995 through the fall of 1998, and the second one, which ran from the late fall of 1998 until the fall of 2000 (see ***Figure 6.7*** on the following page).

In the first phase of dollar strength, the dollar was driven higher by a classic rise in U.S./foreign real interest-rate differentials that made dollar assets more attractive to international investors. As shown in ***Figure 6.8,*** the dollar had been pushed to extreme undervalued levels in 1995, in large part due to an adverse trend in U.S./foreign real interest-rate differentials that occurred between early 1994 and the spring of 1995. But beginning in the late spring of 1995 and continuing thereafter until the fall of 1998, U.S./foreign real interest-rate differentials began gradually to rise in favor of the United States as U.S. economic activity expanded at a faster pace than economic growth abroad, and this helped power the dollar higher over that period. A brief downshift in

FIGURE 6.7 **The Dollar's Uptrend, 1995–2000**

Source: Datastream

FIGURE 6.8 **DM/US$ Exchange Rate and U.S./German Real Interest-Rate Differentials**
(10-Year bond yields less CPI, 1993–1998)

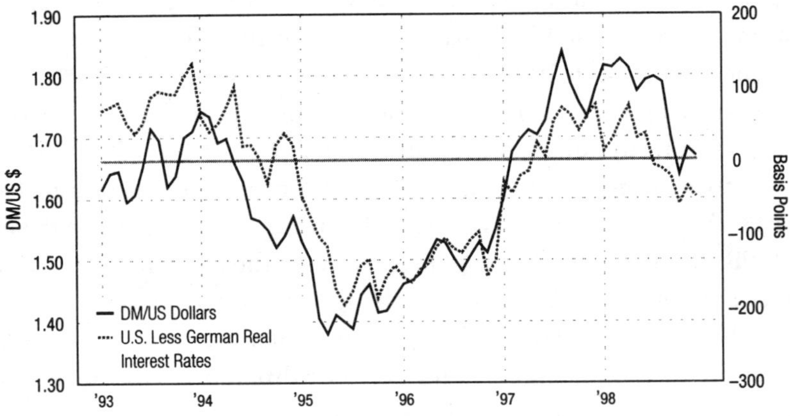

Source: Datastream

U.S./foreign real yield spreads in the fall of 1998 following the Russian/Long-Term Capital Management (LTCM) crisis led to a temporary setback for the dollar, but it was able to regain its lost footing in early 1999 when real rate spreads began to move back in favor of the United States.

Rising real yield spreads in favor of the United States helped attract capital inflow to the United States at an accelerating pace in the second half of the 1990s. In fact, those inflows came in faster than the rate at which the U.S. trade and current account balances were deteriorating. Hence, overall balance of payments forces were highly dollar positive over the 1995–1998 period.

This is evident in ***Figure 6.9***, where we show how a modified measure of the U.S. basic balance of payments (i.e., net private foreign purchases of U.S. equities, agency bonds and corporate bonds minus the underlying trend in the U.S. trade balance) began to rise steadily over the spring 1995–fall 1998 period. As this occurred, the dollar rose in tandem.

The second phase of dollar strength, which began in late 1998/early 1999 and then carried forward until the fall of 2000, was an entirely different animal. Indeed, as shown in ***Figure 6.10*** on the following page, the trend in U.S./foreign real yield spreads argued for a weaker dollar in 1999–2000, not a stronger dollar. Yet the dollar soared over this period.

So, what drove the dollar higher over the 1999–2000 period? We believe the move was propelled by an upward revision in the market's assessment of the dollar's real long-run equilibrium value. This reassessment was sudden and dramatic, following closely on the heels of a sudden and dramatic upgrade in market expectations regarding the U.S. economy's long-run growth prospects.

FIGURE 6.9 **Modified U.S. Basic Balance of Payments and the Dollar**
(Total purchases excluding Treasuries)

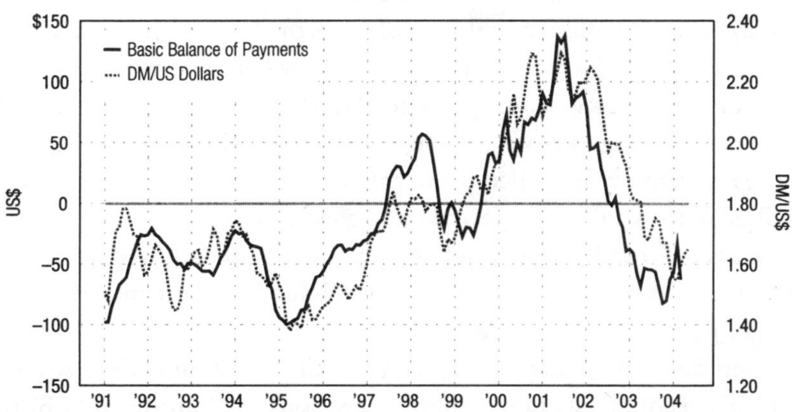

Note: Net foreign purchases of stocks and bonds (excl. Treasuries) less the U.S. trade deficit.

Source: Datastream and U.S. TIC.

FIGURE 6.10 **Modified U.S. Basic Balance of Payments and the Dollar**
(Total purchases excluding Treasuries)

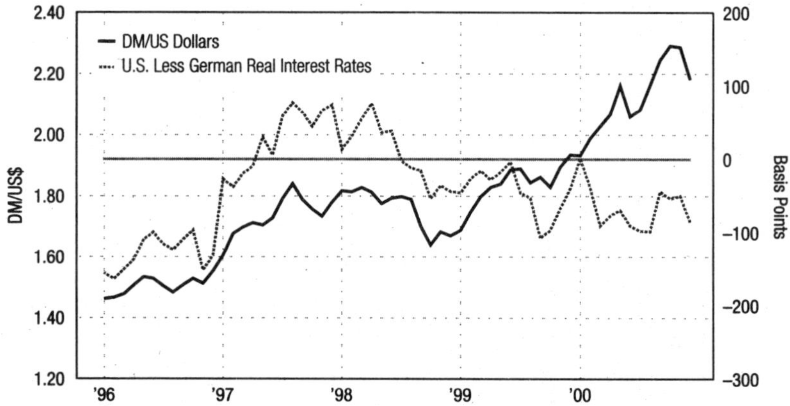

Source: Datastream

Although real GDP growth was beginning to pick up strongly in the United States in the second half of the 1990s, economists initially believed that those gains would prove to be temporary. Their predictions in the second half of the 1990s indicated that they saw little reason why the long-run trajectory for U.S. economic growth needed to be revised upward. This is evident in ***Figures 6.11*** and ***6.12***, which show projections for long-term U.S. productivity and GDP growth during the 1990s, as reported by the Federal Reserve Bank of Philadelphia (whose annual survey of professional forecasters asks economists each February: What annual rate of growth do you expect U.S. real GDP and productivity to average over the next ten years?). As shown, the projected average long-term real GDP and productivity growth rates barely changed from one year to the next in the second half of the 1990s. Each year, polled economists projected that long-term U.S. GDP growth would average around 2.5 percent per annum for the ten years that followed, a fairly modest pace, while long-term productivity growth was expected to rise at a fairly subdued 1.5 percent per annum average pace.

Then something happened between the 1999 and 2000 surveys. Suddenly, economists raised their projections for long-term U.S. real GDP growth from 2.5 percent per annum to over 3.0 percent, and at the same time raised their estimates for long-term U.S. productivity growth from a bland 1.5 percent per annum pace to a brisk 2.5 percent.

FIGURE 6.11 **How Long-Term Forecasts of U.S. Productivity Growth Have Changed in the Past Decade**

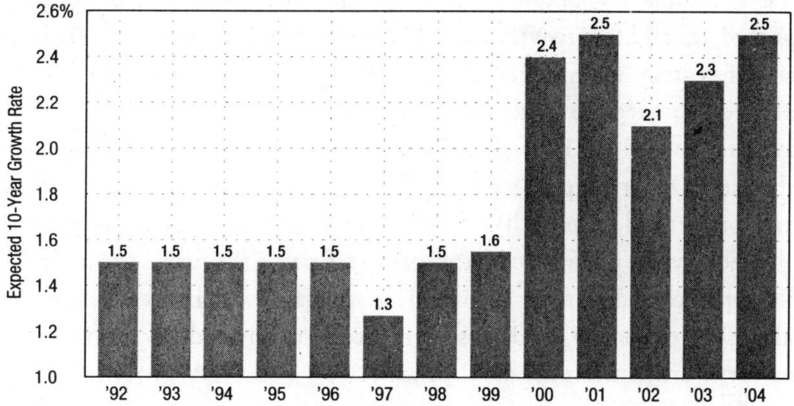

Source: Federal Reserve Bank of Philadelphia Survey of Professional Forecasters

FIGURE 6.12 **How Long-Term Forecasts of U.S. Real GDP Growth Have Changed in the Past Decade**

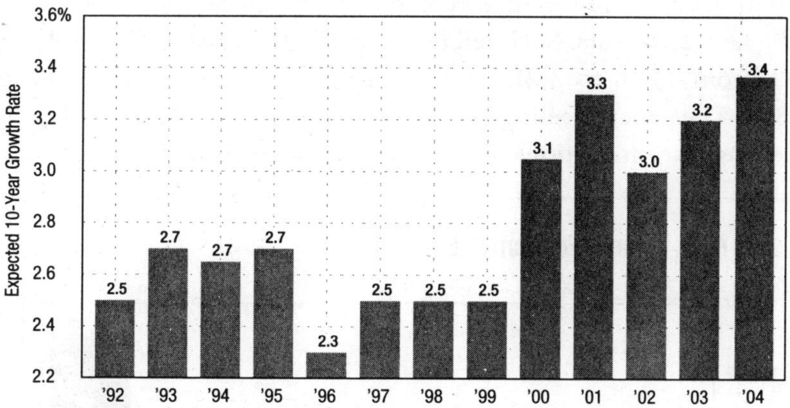

Source: Federal Reserve Bank of Philadelphia Survey of Professional Forecasters

Normally, one would have expected any changes in the U.S. long-term growth outlook to have taken place gradually over a number of years, and not suddenly. But that was not the case this time around. Instead, it appears that expectations regarding U.S. long-term growth prospects not only soared, but this change occurred virtually overnight.

What is interesting about Figures 6.11 and 6.12 is that although U.S. real GDP and productivity growth were advancing quite sharply in the second half of the 1990s, professional forecasters apparently had not yet embraced "new economy" notions at the outset of the IT revolution. Thus the strong pickup in U.S. growth in the late 1990s was perceived to be a temporary phenomenon; in the long run, it was generally expected that U.S. growth would settle back to a more modest pace. That, of course, did not happen, as U.S. growth year in and year out continued to surprise analysts. As a result, the latter half of the 1990s proved to be one of the worst periods ever for professional forecasters, with one of their longest strings of consecutive large negative forecast errors on record. Quite likely, with U.S. growth persistently above projected levels, professional forecasters finally threw in the towel in 2000 and conceded that "something special" was indeed happening to the U.S. economy, and this warranted an upward revision in their long-term projections for U.S. productivity and GDP growth.

Professional forecasters were not the only observers who were late in recognizing that "something special" was taking place on the U.S. economic front. References to the term "new economy" in the U.S. business press were not all that frequent in 1995–1997 (see ***Figure 6.13***), and while references did double in 1998, they paled in comparison to the number of references that were to come in the new economy frenzy of the next two years. Such references in the financial media tripled in 1999, from 1,048 to 3,215, and then soared seven-fold in 2000, to 22,848. The general feeling among economists and market participants alike was that innovations in information technology were beginning

FIGURE 6.13 "New Economy" Euphoria

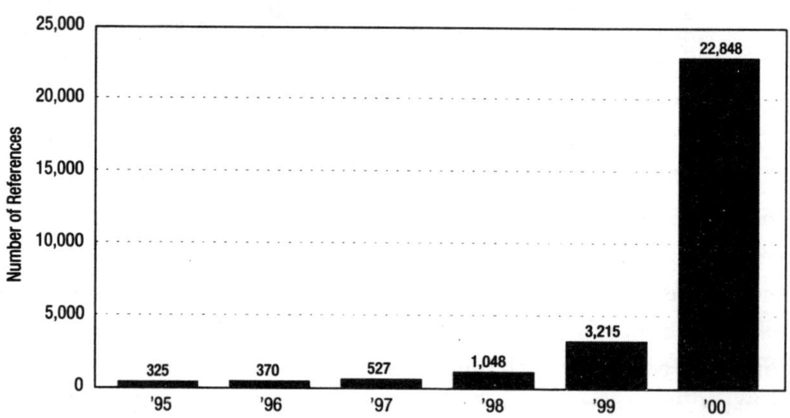

Source: Federal Reserve Bank of Philadelphia Survey of Professional Forecasters

to change the way business firms could add value, and that U.S. firms in particular were at the forefront of those changes on a global scale.

In our view this "new economy" euphoria was the principal driving force behind both the dramatic rise in the Nasdaq index between late 1998 and the first quarter of 2000 and the surge in the dollar's value over the same period. Indeed, the dollar rose sharply in 1999–2000 as capital inflows to the United States surged. With foreign investors eager to participate in the U.S. "new economy" boom, the pace at which overseas capital flowed into the United States far exceeded the pace at which the U.S. trade deficit was deteriorating during this period. This is evident in Figure 6.9, where the United States modified basic balance-of-payments indicator surged in 1999 and continued to advance in 2000 and early 2001. With the dollar largely moving in sync with the improving U.S. basic balance of payments, it was able to appreciate significantly in 1999–2000, even without relative interest-rate support.

As stated earlier, the dollar's dramatic rise pushed it into overvalued territory on a PPP basis. This raised an important question for policymakers and market participants alike: How much of the dollar's rise over this period was an equilibrium phenomenon, and how much of it was a disequilibrium phenomenon? A strong case can be made that the upward adjustment in the U.S. long-term GDP and productivity growth outlook should have resulted in some upward adjustment in the dollar's real long-run equilibrium value, and perhaps to an upward adjustment in the market's notion of what constituted a sustainable current-account deficit for the United States. But by how much?

Figure 6.14 addresses this question on a theoretical level: it illustrates how the IMF estimates a currency's real long-run equilibrium value. As drawn, the IMF framework assumes that the dollar's equilibrium value is determined at the point where the underlying U.S. current-account balance just matches the sustainable inflow of capital to the United States at q_1. Most economists have estimated that the sustainable capital inflow into the United States historically has been around 2.5 percent of U.S. GDP. Given this assumed steady inflow of capital, the United States should therefore be able to safely run current-account deficits of around 2.5 percent of GDP each year without placing undue downward pressure on the dollar's value.

Now consider the possibility that the U.S. economy undergoes a major investment boom that results in a permanently wider gap

FIGURE 6.14 "New Economy" Forces and the Rise in the U.S. Sustainable Current-Account Deficit

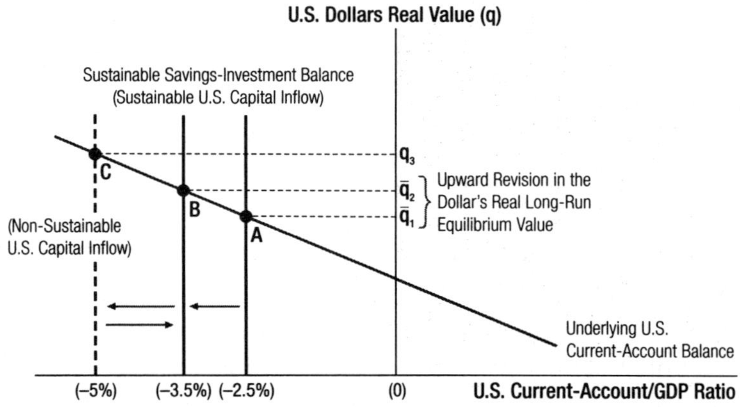

Source: M. Rosenberg, Deutsche Bank

between U.S. investment and savings. That would be possible only if the United States were able to attract additional inflows of capital from abroad on a sustained basis to finance the wider gap between domestic investment and savings. That, in turn, would be possible only if the expected rate of return on U.S. assets were sufficiently attractive to induce overseas capital to flow to the United States on a permanent or semipermanent basis.

So, if the United States could now attract more capital inflow on a sustained basis, then it could safely run larger current-account deficits on a sustained basis than was previously the case. How much did the sustainable inflow of capital to the U.S. rise as a result of the "new economy" boom? It is difficult to come up with a precise estimate, but if the long-term trend in U.S. real GDP growth had risen from, say, 2.5 percent to around 3.5 percent per annum, then perhaps the sustainable inflow of capital to the U.S. might have risen from its previous level of 2.5 percent of GDP to around 3.5 percent of GDP today.

Figure 6.14 illustrates how a rise in the sustainable capital inflow to the United States—from 2.5 percent to 3.5 percent of GDP—would have led to a similar rise in the sustainable current-account deficit that the United States could safely run from 2.5 percent to 3.5 percent of GDP. Note that a rise in the sustainable capital inflow to the United States would also be expected to generate significant upward pressure on the dollar's value on a permanent or semipermanent basis, shown in Figure 6.14 as an upward adjustment in the dollar's long-run equilibrium value from q_1 to q_2.

The issue for the foreign exchange (FX) markets was whether the dollar's actual value might have risen beyond q_2 to, say, q_3 during its 1999–2000 run-up. The evidence suggests that was probably the case, since the U.S. current-account deficit not only widened beyond "old economy" estimates of current-account sustainability (2.5 percent of GDP) but eventually surpassed even the wider sustainability projections based on "new economy" notions (3.5 percent of GDP).

Indeed, while a case could be made that "new economy" forces allowed the U.S. to attract more capital inflow from abroad, in turn allowing the nation to run larger current-account deficits on a sustained basis than before, that did not mean that the U.S. could run any size deficit that it wished. If, for example, the sustainable U.S. current-account deficit had risen from 2.5 percent of GDP to, say, 3.5 percent of GDP, then a deficit that exceeded this new upper limit would need to be eliminated, either through a decline of the dollar, a major slowdown in U.S. GDP growth, or some combination of the two.

As a matter of fact, the dollar's rise over the 1995–2000 period began to meet resistance in late 2000 when the U.S. current-account-deficit/GDP ratio began to rise above the 3.5 percent threshold level (see ***Figure 6.15***). The dollar did not immediately decline in value when this threshold level was initially breached, because relatively

FIGURE 6.15 **U.S. Current-Account Balance, 1980–2004**

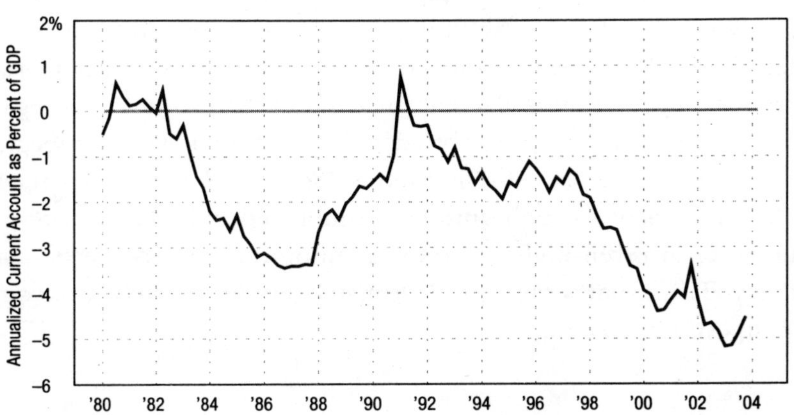

Source: Datastream

FIGURE 6.16 **2000 Nominal Short-Term Interest Rates**
(Three-month Euro-deposit rates)

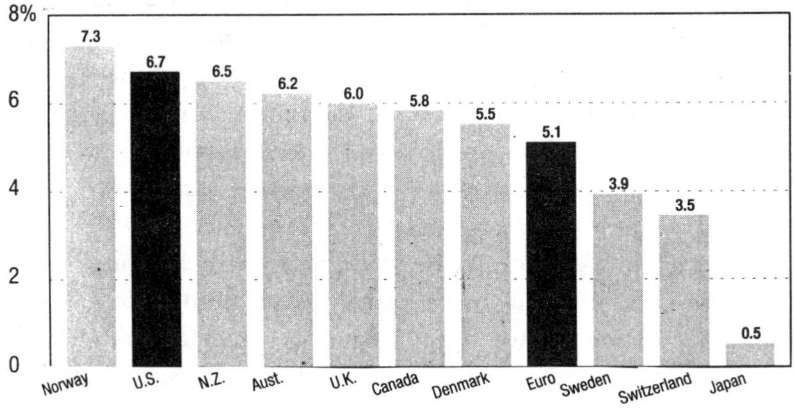

Note: As of October 2000

Source: Datastream

high U.S. interest rates in 2000 and early 2001 continued to attract the necessary capital inflow to finance the widening U.S. current-account shortfall. As shown in ***Figure 6.16***, up until late 2000 U.S. short-term interest rates were among the highest in the industrial world. That changed as 2001 rolled along, however. With the U.S. economy drifting into recession in 2001, the Federal Reserve aggressively pushed U.S. short-term interest rates lower, and by 2002, U.S. short-term rates had moved from being among the highest to being among the lowest in the industrial world (see ***Figure 6.17*** at right).

This sharp decline in U.S. short-term interest rates set the stage for the inevitable decline in the dollar, which began in earnest in early 2002. The U.S. current-account deficit had always been a potential source of danger for the dollar, and once short-term interest rates began to fall in 2001, the risk/reward tradeoff of holding dollar assets began to tilt against the dollar. As illustrated in ***Figure 6.18***, the United States shifted from being a high-current-account-deficit/high-interest-rate country to an even-higher-current-account-deficit/low-interest-rate country. This major regime shift ultimately brought the dollar's long-term upcycle to an end.

FIGURE 6.17 **2002 Nominal Short-Term Interest Rates**
(Three-month Euro-deposit rates)

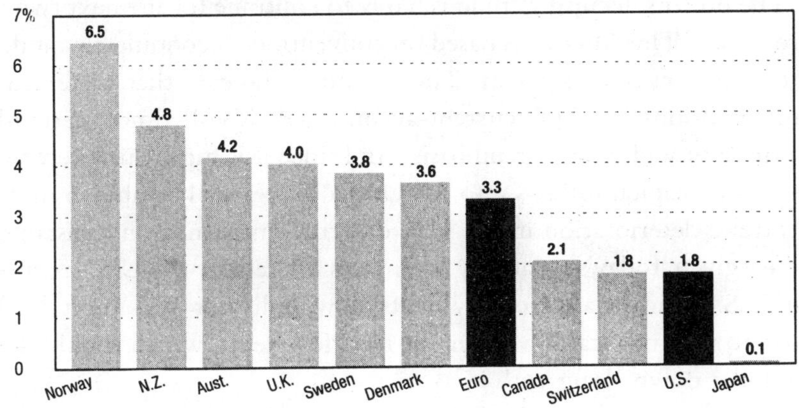

Note: As of October 2000

Source: Datastream

FIGURE 6.18 **Interest Rate Current Account Nexus**
(Total purchases excluding Treasuries)

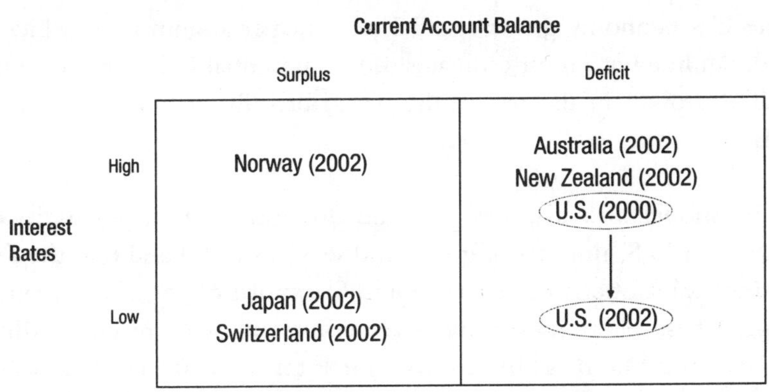

Source: M. Rosenberg, Deutsche Bank

The U.S. Current-Account Deficit and the Dollar

The year 2001 marked a watershed for the dollar. As U.S. short-term interest rates drifted to new cyclical lows, net private foreign purchases of U.S. securities began to fall off in the second half of 2001. With the

U.S. trade deficit continuing to widen to new record levels, the U.S. basic balance of payments began to deteriorate, and the dollar fell in tandem (see Figure 6.9).

The dollar's declining trend is likely to continue for the next two to three years. This forecast is based on conventional econometric models of the U.S. current account. Those models indicate that under current conditions, the U.S. current-account deficit will almost certainly continue to widen on a trend basis. And since it is highly unlikely that net private capital inflows into the United States will be able to match this trend deterioration in the U.S. external imbalance on a sustained basis—given today's relatively low rates of return on U.S. assets—our U.S. basic balance-of-payments flow indicator will most likely move to new cyclical lows over the next few years, which should drag the dollar down along with it.

Using a conventional econometric model of the U.S. external imbalance, we project that the U.S. current-account deficit will widen from roughly 6 percent of GDP today to 8 percent by 2008, assuming no change in the dollars value and no change in U.S.–foreign GDP growth differentials (see ***Figure 6.19*** at right). More specifically, our baseline projection incorporates the following assumptions:

1. the dollar remains at its present level;
2. the U.S. economy grows at a 3.5 percent per annum pace, which is broadly in line with most estimates of U.S. potential GDP growth; and
3. GDP growth in the rest of the world broadly matches the United States.

The model also assumes that the domestic income elasticity of demand for U.S. imports of goods and services is 2.0 and that the foreign income elasticity of demand for U.S. exports of goods and services is 1.4. (These elasticity estimates are consistent with recent findings by Catherine Mann of the Institute for International Economics. See "The U.S. Current Account, New Economy Services, and Implications for Sustainability," January 2004, which is available on the IIE.com website.) The higher income elasticity of demand for imports relative to exports, coupled with the fact that the present level of U.S. imports is more than 50 percent larger than U.S. exports, virtually guarantees that there will be a trend widening in the gap between U.S. imports and exports over time unless this is offset by slower U.S. growth and/or a major slide in the dollar's value.

FIGURE 6.19 **Deutsche Bank's U.S. Current-Account Balance Model Baseline Projections**

(Annualized current-account balance)

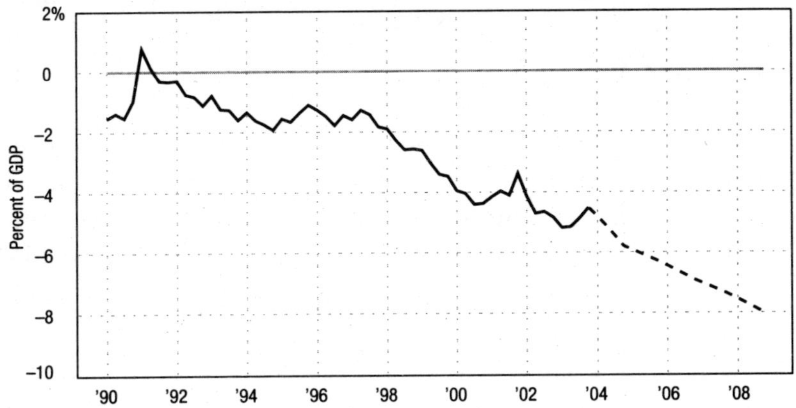

Source: Datastream and Deutsche Bank's projections

Although the baseline assumptions point to a widening of the U.S. current-account deficit from 6 percent to 8 percent of GDP, it is almost inconceivable that the markets would allow the deficit to reach such an outsized level. A recent study by Caroline Freund of the Federal Reserve indicates that the threshold level for current-account deficits in industrial countries is about 5 percent of GDP. When deficits have exceeded that level, corrective forces have often come into play—in the form of slower domestic GDP growth and major currency depreciation—to restore the current-account deficits to sustainable levels.

Our econometric model can be used to determine how much the U.S. economy might need to slow and how much further the dollar might need to fall in order to restore the U.S. current-account balance to a sustainable level. As shown in ***Figure 6.20***, if the dollar were to remain at present levels and at the same time the U.S. economy were to slow from a 3.5 percent growth pace to 2.0 percent per annum on average for the next five years (while the rest of the world's growth rate remained at the same pace as was assumed in the baseline simulation), the U.S. current-account-deficit/GDP ratio should narrow relative to the baseline projected path, but the deficit would probably remain unsustainably wide at around 6 percent of GDP in five years' time.

FIGURE 6.20 **Deutsche Bank's U.S. Current-Account Balance Model**
(Annualized current-account balance)

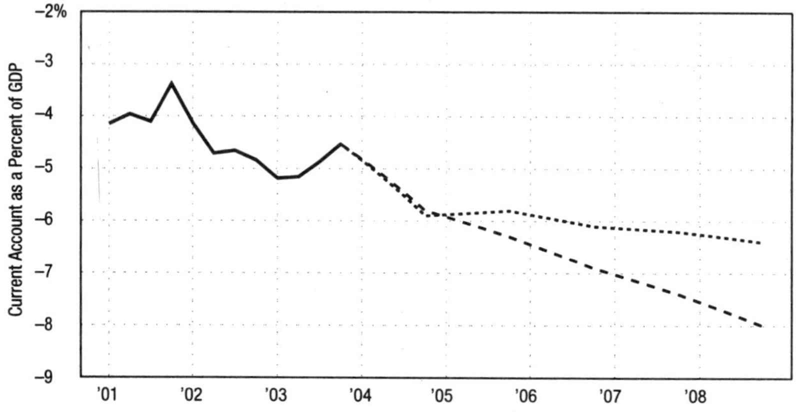

Note: Assuming a sustained slowdown in W.S. GDP growth and a stable dollar in 2004–08

Source: Datastream and Deutsche Bank's projections

What this suggests is that while a sustained slowdown in U.S. GDP growth would assist in the correction of the U.S. external imbalance, it would not be enough to restore the U.S. current-account deficit to a sustainable level. Assuming that a more pronounced slowdown in U.S. GDP growth to below 2.0 percent per annum would be resisted by U.S. policymakers, the only way the U.S. current-account deficit can be reduced further is if the dollar plays a substantial role in the trade-adjustment process.

Figures 6.21 and ***6.22*** provide alternative projections for the U.S. current-account deficit, assuming both a significant slowdown in U.S. GDP growth (to 2.0 percent per annum on average over the next five years) and a sharp drop in the dollar's value. Figure 6.21 assumes that the dollar's trade-weighted value falls by 10 percent in 2004 and another 10 percent in 2005. The cumulative 20 percent depreciation of the dollar, in conjunction with slower U.S. GDP growth, should help bring the U.S. current-account deficit/GDP ratio down to around 4 percent of GDP by 2008. Figure 6.22 assumes that the dollar's trade-weighted value falls by 10 percent in 2004, another 10 percent in 2005, and still another 10 percent in 2006. The cumulative 30 percent depreciation of the dollar should help narrow the U.S. current-account deficit to around 3 percent of GDP by 2008.

FIGURE 6.21 **Deutsche Bank's U.S. Current-Account Balance Model**

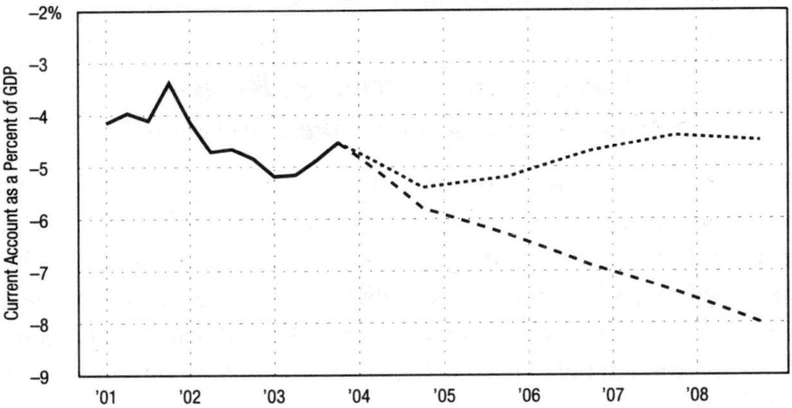

Note: Assuming a sustained slowdown in U.S. GDP growth and a 20% decline in the dollar in 2004–05

Source: Datastream and Deutsche Bank's projections

FIGURE 6.22 **Deutsche Bank's U.S. Current-Account Balance Model**

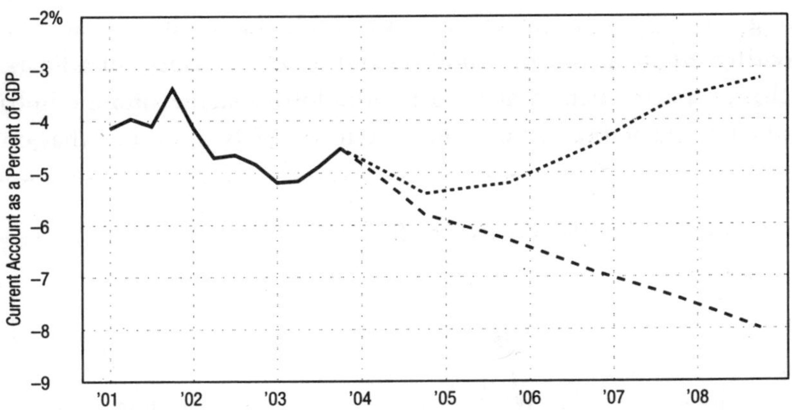

Note: Assuming a sustained slowdown in U.S. GDP growth and a 30% decline in the dollar in 2004–06

Source: Datastream and Deutsche Bank's projections

Whether the dollar will need to fall by 20 percent or 30 percent, or possibly even more, will ultimately depend on what U.S. deficit/GDP ratio is deemed to be sustainable in the long run. If this ratio is closer to 2.5 percent to 3.0 percent of GDP, then, according to our simulations, the dollar might need to fall by at least 30 percent over the next few years. If, instead, one embraces a "new economy" view that the

U.S. sustainable current-account deficit has now been raised to 3.5 percent to 4.0 percent of GDP, then the dollar might have to fall only by around 20 percent over the next few years.

Overshooting Exchange Rates— How Should Policymakers Respond?

A further 20 percent to 30 percent decline in the dollar's value over the next two to three years would constitute a major downside overshoot for the dollar on a PPP basis. Many market commentators and policymakers will probably view such a move as excessive and will therefore recommend that efforts be taken to resist it. But is it really in the interests of U.S. and foreign policymakers to prevent the dollar from overshooting to the downside?

Exchange-rate overshooting is not an uncommon phenomenon. As shown in ***Figure 6.23***, although the dollar typically fluctuates within a +/– 20 percent range around its PPP level, roughly every five to seven years or so, the dollar has tended to overshoot this range on both the upside and downside by rather large amounts. This raises a set of interesting and important questions: Is it possible that exchange-rate overshooting might in fact be a normal and necessary aspect of a floating exchange-rate system to help deal with global macroeconomic imbalances that inevitably occur from time to time? Is it possible that only

FIGURE 6.23 **U.S. Dollar Purchasing Power Parity**

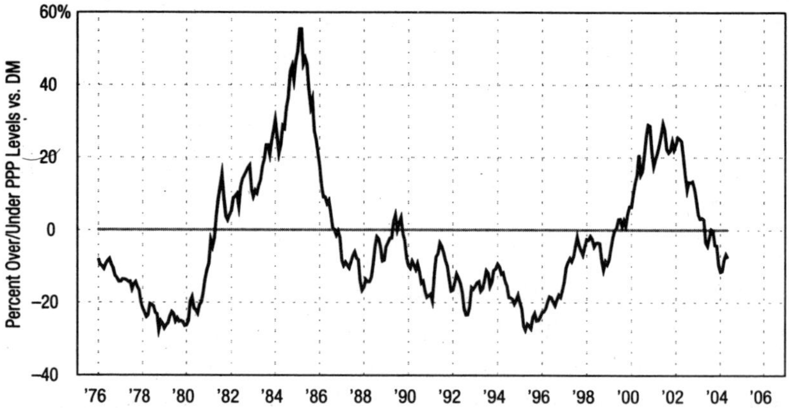

Note: PPP estimates based on 1982–2000 averages

Source: Datastream

very large changes in exchange rates are capable of correcting large macroeconomic imbalances?

Consider the role that exchange rates typically play in a country's trade-adjustment process. Most empirical studies find that the pass-through effect of exchange-rate changes on U.S. traded-goods prices is limited. Even then, once traded-goods prices do shift in response to an exchange-rate change, the evidence indicates that the price elasticity of demand for traded goods is typically low, suggesting that there will be only a muted response in U.S. and foreign demand to a change in traded-goods prices. Therefore, given limited pass-through effects and low price elasticities of demand, it is unlikely that modest changes in exchange rates can successfully bring about significant changes in trade imbalances. Rather, it appears that only very large exchange-rate movements are capable of bringing about the changes in trade volumes necessary to correct large external imbalances.

This explains why our econometric model of the U.S. current-account balance shows that only a very sharp decline in the dollar's value can restore the U.S. external imbalance to a more sustainable level. However, if policymakers attempt to limit the downside move in the dollar that our econometric model is calling for, then such actions would impede the adjustment of the U.S. trade imbalance to a sustainable level. That would not only push back the inevitable U.S. trade adjustment to a later date, it might eventually lead to a more dramatic decline in the dollar in the long run if the world financial markets have to cope with even larger, more intractable U.S. trade imbalances in the future.

Although foreign policymakers might not want to see their currencies rise too sharply against the U.S. dollar, we believe it is in their long-term interests to see the U.S. current-account deficit narrow to a sustainable level. Historically, large current-account imbalances have acted as a constraint on individual countries' long-term growth prospects. Thus, it is not in the interest of the United States or the rest of the world to see the U.S. economy's long-run growth prospects compromised. After all, the United States has been and is likely to continue to be one of the principal growth engines of the world economy. The world economy's long-term growth prospects are likely to be enhanced if policymakers simply stand out of the way as the market gropes to find a value for the dollar that will help restore the U.S. current-account balance to a sustainable level in the long run.

7

Politics and the Economy

THOMAS D. GALLAGHER

THOMAS GALLAGHER links economics and politics at Ed Hyman and Nancy Lazar's International Strategy & Investment Group, Inc. Gallagher casts an experienced, skeptical eye, giving political analysis that makes him the first call for ISI's clients, politicians, and the media. Gallagher's secret weapon is rock-solid economic analysis combined with that rare Washington skill—he listens carefully. Here, Thomas Gallagher takes the pulse of politicians as they debate near-term, vote-friendly economics and postpone difficult decisions for the political economics of our children.

Here is a premise: Second presidential terms in recent history have been disappointing, but George W. Bush has an opportunity to avoid that fate. The last three reelected presidents (Bill Clinton, Ronald Reagan, and Richard Nixon) all saw at least one chamber of Congress controlled by the opposition party. With both houses now controlled by Republicans, President Bush's situation is more comparable to those of LBJ and FDR in that he has a much greater opportunity to implement his agenda. This chapter looks at three standard aspects of political economy: macroeconomic policy, the role of government in the economy, and international economic policy. The last part examines the policy impact of increased polarization in Congress.

Countercyclical Policy— Changing from Stimulus to Restraint

In 2004, economic conditions changed, so the imperative for macroeconomic policy shifted from stimulus to restraint. Institutionally it has been much easier for monetary policy to swing to restraint than it has been for fiscal policy, suggesting a policy mix of tighter money and easier fiscal policy for the near term. This policy mix should have implications for interest rates, investment, and the dynamics of any adjustment of the current account deficit.

A dominant feature of the U.S. political economy since 2001 has been massive macroeconomic stimulus. The real fed funds rate hovered around zero or below for three years. Fiscal stimulus by some measures has exceeded even the tax cuts and defense buildup of the early Reagan years. With a fairly broad consensus that the recovery is sustainable, this degree of stimulus was no longer appropriate. Thus, the key theme for macroeconomic policy has been the swing from stimulus to restraint.

An important question that will frame the dynamics of this swing to restraint is the nature of the last recession. If this was just a normal recession made worse by shocks (terrorism, war-related risk aversion, and corporate scandals), this stage of the business cycle will be similar to previous recoveries. To the extent this was a cycle driven by post-bubble drags, imbalances that remain to be worked out may moderate the degree of restraint that policy must deliver. That is, unlike previous recoveries that were characterized by unleashing pent-up consumer demand, this time household spending may be quite restrained due to the heavy buildup of debt over the past few years.

Monetary Policy

Due to the independent structure of the Federal Reserve Board, there are no institutional barriers to monetary policy swinging from stimulus to restraint. It seems likely that the magnitude of tightening by the Fed will be held down due to several headwinds on the economy—caution in business spending, a low saving rate, and the lagged effect of higher oil prices.

Institutionally the most important political development for the Fed near term will be President Bush's choice of Alan Greenspan's successor as Fed chairman. Although his term as chairman was extended in 2004, Greenspan will almost certainly be replaced in the first half of 2006. His term as a governor, as opposed to chairman, expires in

January 2006, and he's not eligible for another term (Fed governors who have served a full term can't be appointed to another one). The law allows governors whose terms have expired to continue serving until a successor has been confirmed, but this backdoor extension of Greenspan's term isn't likely to be employed for long, since it would be seen as an untenable challenge to the independence of the Fed to have the chairman in effect serving at the pleasure of the president and Congress. He could serve a few months after January (he would become the longest serving Fed chairman ever in June 2006), so it's very likely there will be a new Fed chairman by mid-2006.

Two things should be said about this transition. One is that the Fed will probably become more inflation-vigilant in the run-up to the change. The last thing the Fed would want is rising inflation or inflation expectations just as the person who embodies anti-inflation credibility is about to depart. The other is that there will be more uncertainty about expected short-term interest rates starting in late 2005. One accomplishment under Greenspan has been greater transparency, and financial markets will be less clear about the Fed's reaction function under a new chairman. Will the new chairman be as dominant a figure in setting monetary policy as Greenspan, and if so how will he feel about "measured" tightening, the role of central banks during asset bubbles, and productivity trends, for example?

Fiscal Policy

As easy as it has been for monetary policy to swing from stimulus to restraint in the face of a sustainable recovery, that is how hard it will be for fiscal policy to follow suit. In traditional countercyclical policy, stimulus is withdrawn as the recovery takes hold. The impact of fiscal policy on GDP growth is measured by the year-over-year change in the budget deficit (adjusted for cyclical changes in the economy). On that measure, fiscal policy will turn slightly restrictive in 2005, due to the expiration of bonus depreciation. But the level of stimulus will remain fairly high, with bipartisan support to extend the middle-class tax cuts that were scheduled to expire at the end of 2004 but were extended just before the election.

The more fiscal stimulus is withdrawn, the less work monetary policy has to do. Put another way, a failure to withdraw stimulus will lead to a somewhat higher interest rate path for the economy over the next few years. So, what are the prospects for reduced fiscal stimulus? The prospects are poor. Notably after the last two spikes in the budget deficit—in the early 1980s and the early 1990s—there was serious bipartisan interest in paring back the deficit. That kind of bipartisan interest isn't at all evident in 2004.

Several factors argue for low expectations—the lack of recent experience with fiscal restraint, the political costs imposed on major efforts at fiscal restraint in the 1990s, the polarized nature of Congress, the worry by small-government Republicans that engaging in deficit reduction could lead to tax increases, and the lack of either presidential candidate having sought a mandate to rein in stimulus all point to stalemate or minimal progress in whittling away stimulus.

It has been years since Washington grappled with fiscal restraint. Despite the apparent economic success of budget surpluses, the political history of the 1990s doesn't show political rewards for specific efforts at fiscal restraint. The first President Bush owed an important part of his defeat to the tax increases that were part of his 1990 budget package. The tax increases that were the centerpiece of Clinton's 1993 budget contributed to the party's crushing defeat in the 1994 midterm election. It wasn't just tax increases that were unpopular. The Republican Congress used government shutdowns as leverage to implement major spending restraint, including Medicare savings, which shaped the presidential election in 1996. The last time fiscal restraint was addressed in a meaningful way was in 1997, when President Clinton and the GOP-led Congress reached a balanced budget agreement. But by then budget projections had become favorable enough that only modest restraint was needed to achieve a balanced budget. In fact, the restraint in the 1997 deal is best seen as a small toll that allowed each side to claim victory on a balanced budget and to address its own fiscal initiatives (spending for Democrats, tax cuts for Republicans). After that, the fiscal focus shifted to disbursing surpluses and then to fighting the recession with stimulus.

For several reasons, the current President Bush is likely to seek to reverse the spending growth trends of his first term—the revolt among small-government conservatives (especially after the drug benefit's enactment), concerns about the financial market reaction (in both the bond and currency markets) to rising deficits, greater congressional majorities, and the fading of the emergencies that led to the spending surge in the first place. But the impact of this newfound interest in spending restraint will be muted by the long-term deficit impact of Social Security reform and making the tax cuts permanent.

The Return of Big Government

Perhaps the most surprising feature of the U.S. political economy of the past few years is the expansion of government in both economic and non-economic spheres—surprising because for most of the time an all-Republican government was in place.

Consider what followed the Republican sweep of the midterm elections in 2002. The first action in the lame duck session was for the party to enact legislation creating a new Department of Homeland Security, the first new cabinet department since 1989. One of the first actions in the new Congress in 2003 was an extension of unemployment benefits. And a top domestic priority of President Bush and the Republican leadership was the biggest expansion of the welfare state in years—the enactment of a new drug benefit under Medicare.

By just about any measure or from any perspective, government's role expanded. This trend had several causes:

9/11 and the war on terrorism—Throughout history wartime has produced growth in government. This produced rising defense and homeland security spending, along with new interventions in the economy (terrorism insurance, the airline bailout, and the transfer of airline security workers from the private to the public sector). It also caused the needle to swing from civil liberties toward police powers of the state.

The recession and post-bubble drags on the economy—At the micro level, the accounting, finance, and corporate scandals associated with the tech bubble produced a range of new regulations, typified, but certainly not limited to, the Sarbanes-Oxley legislation (2002), which imposed stricter governance rules on business and auditing. At a macro level, the recession and extended jobless recovery provoked not just the expected monetary easing but substantial fiscal activism as well.

Aging of the population—Catering to aging baby boomers produced the biggest expansion of an entitlement program in decades. Enactment of the drug benefit in 2003 was an implicit acknowledgement that the government's role should be expanded to include restraining the cost of prescription drugs.

Longer-term political trends—It may well be that the drive toward smaller government that began in the late 1970s, but accelerated under President Reagan, has largely spent its course. The drive for deregulation has waned, as shown by the large congressional majorities that supported overturning the lifting of the ownership rules for television stations. It may be that the combined effect of the special factors mentioned above has been to generate more support for the kind of collective efforts that government provides, which in turn has robbed the drive for scaling back government of any political momentum. The fact that nondefense, non-homeland security spending has also outpaced nominal GDP growth suggests that support for a bigger government isn't just tied to the traditional security function that is the classic public good.

In contrast, it's hard to find any political successes on the smaller government agenda. The only argument that has been advanced is the "starve the beast" argument for tax cuts. But the drug benefit is pretty solid evidence against that rationale. The ink was barely dry on the tax cut when Congress, in the face of the largest deficit in history, enacted the new Medicare benefit that was financed mainly by borrowing. The beast doesn't look very hungry.

What does the future hold? It is fair to say that Republicans are in power *despite* being the party of smaller government, not because they are the party of smaller government. Most analyses of the 2004 election suggest that it was cultural and security issues that gave Republicans their hold on power. Nonetheless, winners get to act on their entire agenda, so it's reasonable to think Republicans will try to roll back government's role in the economy somewhat. Here are some issues on which the "size of government" agenda will be played out.

Ownership Society and Social Security reform—Small-government conservatives pin their hopes on President Bush's "ownership society" notion as their best way to shrink government. The argument is that shrinking the supply of government hasn't been successful, so shrinking the demand for it may be the better strategy. Many government programs essentially insure against various risks. For example Social Security insures against income loss due to retirement and disability. The way to reduce dependence on the government for such insurance is to encourage individuals to save for these purposes. That's a unifying theme behind Bush's proposal for Social Security private accounts, for Health Savings Accounts, and for Personal Reemployment Accounts. Thus although the president avoids the term "privatization," his proposal for personal accounts would essentially be a step toward privatizing the retirement income security function that the government now serves and would represent a major victory for proponents of smaller government.

Whatever progress may be made on improving Social Security's long-term financial situation has to be put in the context of the damage done by the Medicare drug benefit. The increase in Medicare's fiscal imbalance (measured as the present value of future unfunded liabilities divided by the present value of future GDP) caused by the drug benefit is greater than the entire fiscal imbalance of the Social Security program. So even if Social Security's long-term problems are completely fixed in the next few years, that wouldn't reverse the deterioration in long-term finances caused by the drug benefit.

Near-term deficit reduction—In President Bush's FY2006 budget, he proposes a slowdown in total spending growth. In his first term annual outlay growth averaged about 7 percent, while his proposal for 2006 is 5 percent. That is accomplished by meaningful restraint in nondefense, non-homeland security discretionary spending (about one-fifth of total spending). But increases outside that spending category keep spending from falling meaningfully as a share of GDP, and new tax cut proposals mean that the only deficit reduction over the next few years comes from the expansion of the economy; there is no net deficit reduction from policy actions.

Health care—Here is a major battleground to expand the role of government. Issues include:

- *Efforts to rein in drug prices*—Drug companies continue to face bipartisan efforts to lower drug prices by, for example, lifting the ban on reimportation of drugs from countries with price controls and making use of generic substitutes more widespread.
- *Revisions to the drug benefit*—The drug benefit was the low-hanging fruit for expanding entitlements, and no other major efforts are obvious. But tinkering with the benefit could provide further opportunities for an expanded government role, such as a more generous benefit financed by having the government use its monopsony buying power in effect to implement price controls.
- *Dealing with the uninsured*—Both parties accept some government role here but differ on how it should be accomplished. Bush would like to implement a tax credit for the uninsured so that they can purchase their own insurance.

Expanding the size of the military—The Iraq war has demonstrated obvious manpower problems, as evidenced by the extended troop rotations, the reliance on National Guards and reservists, the transfer of troops from South Korea, and the debate over the need for a draft.

There's a good chance now that Bush since his reelection will have to expand the size of the military, with certain implications for labor market tightness and wage inflation.

Political Economy of Globalization

There are three overlapping issues regarding international economic policy.

First, in the international arena, will there be support among governments for policies that sustain globalization? Among the major threats to this consensus are disagreements over the appropriate policies to cope with the U.S. current account deficit and foreign policy disagreements, as manifest, for example, in the debate over the Iraq war.

Free trade is not a self-generating policy. Standard political economic analysis argues that free trade is an unnatural political act for legislators (the benefits are diffused, while the costs are highly concentrated). Free trade has needed a strong political underpinning to move ahead, and the Cold War provided that rationale for fifty years. The prosperity of the West was a compelling rationale for its leaders; this meant that heads of government told their lead trade officials to keep trade problems out of their in-box—strike free trade agreements and resolve trade disputes to keep them from interfering with what heads of government cared about most, preserving the Western alliance. In another sense, the gains from cementing the alliance meant that the United States was willing to accept more of the concessions, or absorb the costs, of trade-liberalizing negotiations.

Disagreements over the proper policy toward Iraq are only one of many explanations, but it's probably no coincidence that global free trade efforts largely came to a halt in the past few years. That is, heads of government had little reason to intervene in the stalled Doha round of WTO negotiations, and certainly the United States had little reason to accept a disproportionate share of the concessions in advancing free trade. Resolving trade disputes has also become harder, as evidenced by the increased inability of the United States and the EU to resolve disputes short of taking retaliatory actions. The lack of an underlying security relationship is an important difference in U.S. disputes with China versus those with Japan that characterized much of the 1980s and 1990s.

The risk of a stagnant environment for free trade isn't straightforward. To paraphrase what the late Herbert Stein said about the budget deficit many years ago: the risk isn't that the wolves are at the door but that the termites are in the foundation. Countries are likely to raise trade barriers in this environment. Over time the restraining factor of not wanting to act contrary to the intent and direction of global free trade talks will wane, leading countries to come up with a "Plan B" on trade, which is likely to involve marginally more nationalistic economic policies. Regional approaches to trade and capital liberalization will get more energy and attention.

At the time of this writing there is also disagreement over how to cope with pressures of adjusting to a lower U.S. current account deficit. U.S. policymakers on the one hand and those of its leading trading partners on the other disagree over basic questions. For example, should active currency policies be adopted to limit the dollar's fall, and should the primary burden of adjustment be placed on U.S. fiscal policy, on European and Japanese structural and macroeconomic policies, or on Asian currency revaluation?

Second, in the domestic arena, will there be political support for policies to sustain globalization? This abstract question quickly turns into a practical question: Will the United States impose restrictions on goods trade with China or services trade with India? The best test of trade policy with China will be the lifting of quotas (in the developed world, not just the United States) on textile trade at the end of 2004. This phase-out was agreed to in the Uruguay Round of GATT negotiations in 1994, and nearly everyone's expectation at the time was that by now there would be a debate between the United States and other developed countries to continue some regime of trade restrictions versus developing countries, which would want to stick to the scheduled elimination of textile quotas. But with the emergence of China as the dominant, low-cost supplier of textiles, many developing countries now see a continuation of quotas as in their interest—countries such as Bangladesh (not exactly seen as a high-cost country) think continued quotas may be the only way to preserve their market share in the United States.

The only way that China will face limits is if the U.S. industry successfully uses U.S. trade statutes to win import relief. As the steel industry has demonstrated in the past, one way to generate a regime of import protection is to file many trade complaints, in the hopes of pressuring the U.S. government to negotiate "out of court" settlements with foreign countries in the form of market-sharing arrangements (to avoid the chaotic trade disruptions that could result from these cases). In 2003, the U.S. textile industry filed test cases on knitted fabrics, gloves, and bras (categories that had their quotas lifted in earlier years), and the Chinese have not been willing to negotiate. The precise outcome here is very hard to predict, but even a new regime of limits on Chinese imports would probably produce a substantially greater U.S. market share for Chinese products.

The initial stages of the outsourcing debate offer encouraging news on this front as well. Despite the high level of rhetoric during the past year, the only measure passed by either chamber was a Senate amendment that had so many loopholes and exceptions that it could not meaningfully affect the pace of outsourcing by U.S. firms, and that amendment was dropped from the final bill signed into law. And the Democratic primaries offered one more example that protectionism is not the path to the White House, or even the Democratic Party nomination. Senator John Edwards (D-NC) took a far more protectionist line than did Senator John Kerry (D-MA) and couldn't even win the Ohio Democratic presidential primary. Kerry for all his rhetoric has been a Democratic free trader (having voted for NAFTA and China's WTO membership). His risk in the campaign was less that his outsourcing proposals would interfere with the dynamism of the U.S. economy than it was that the real world impact of his proposals would not match the rhetoric he was using. Without a normal cyclical recovery in jobs, pressures for limits on goods trade with China or outsourcing with India and other countries will continue.

Finally, the challenge of global governance is likely to be a dominant political economic theme for years to come. This is less an explicit issue and more an issue that comes up as part of the backdrop to other issues.

The scope of government action on economic policy tends to match the size of the relevant markets, so economic globalization is increasingly pressing governments to cooperate. In this context, a key consequence of globalized markets is to increase the external ramifications of national policies. Who, in 1997, would have cared about the adequacy of Thailand's banking regulations without freely flowing global capital? What is the World Trade Organization (WTO) dispute-resolution process other than a way to internalize the externalities of national governments' trade policies? The war with Iraq underscores the difficulties of providing that most basic of public goods—security—on a global scale. Suffice it to say that both the multilateral and unilateral approaches to preventing the proliferation of weapons of mass destruction to rogue states and failed states have their drawbacks.

The capacity for global governance will be a major issue. Virtually every supranational institution—the United Nations Security Council, the WTO, the International Monetary Fund, the European Union—faces to differing degrees questions of competence, legitimacy (to national publics), and decision-making structure (driven by a few dominant players or by a more democratic process). Development of adequate institutions will be an erratic and decades-long process that will mean problems of a global scope are less likely to find solutions than problems of a mainly domestic orientation.

Polarized Politics

The increased polarization of Congress has had major implications for the legislative process on economic policies. Increased polarization simply means that Democrats are more liberal and Republicans more conservative.

There are numerous ways to demonstrate increased polarization in Congress. The *National Journal* has published rankings of Senate and House members for more than two decades. In 1999, for the first time it found no ideological overlap between Democrats and Republicans—the most conservative Democrat was more liberal than the most liberal Republican.

Academic studies point to the same conclusion. Various methods rely on comparing individual legislators' voting scores to the averages for their party or their house of Congress. These studies find the two parties' average scores moving farther apart, or find more legislators closer to their party's average than to their chamber's average. Put in a less academic way, there are fewer centrists in Congress.

Why this is happening is less important for the purposes of this chapter than the fact that it is occurring. This trend has had fairly profound impacts on legislating. In earlier periods when there were more centrists, party leaders were chosen on their ability to get things done, and more power resided with committee chairmen. These chairmen were skilled in how to move legislation, since they knew where the various interests groups were, where their committee members were, where the administration was—in short, they knew where the votes were. Bills were usually passed with bipartisan support.

But with fewer centrists, party leaders are now chosen more for their willingness to adhere to an ideological or partisan agenda, and power has shifted away from committee chairman to party caucuses. The result of these trends is a Congress less interested in having trains run on time and more interested in inaction rather than in passing bills with the "wrong" majority. A corollary is that as party caucuses become the relevant arena for decision-making, committee and floor debates become less important, as they become arenas merely to implement decisions made elsewhere.

This leads to three kinds of outcomes in Congress:

Stalemate—A recent academic study found a measure of polarization a statistically significant variable explaining legislative stalemate. It makes sense that in a divided government or divided Congress, consensus is going to be harder to reach.

One-party progress—When there is one-party government, consensus can be achieved. Perhaps the most underreported aspect of President Bush's first term was the productivity of Congress despite the narrowest House and Senate majorities in decades. With increased polarization, party unity is more likely, so either party can govern with quite narrow majorities. Even this is difficult, however, given the need to clear the sixty-vote Senate hurdle to cut off filibusters. So the one-party progress in recent years has usually come under budget reconciliation rules, under which filibusters are not possible. One consequence of that is that changes can't be permanent—consider the sunset dates for the Bush tax cuts, for example.

Harvesting low-hanging fruit—The two parties occasionally can agree on issues that don't strike deeply into the divisions that define the two parties, wherein neither party feels the other side achieves more of an advantage. The Medicare drug benefit is a recent example. Reaching back to the 1990s, examples of such low-hanging fruit are the mid-1996 compromises on a minimum wage increase, health insurance portability, and welfare reform.

What are the implications of these outcomes? To the extent stalemate is more likely, financial markets won't necessarily regard that as a bad outcome. So, citing stalemate as a more likely outcome is just a description, not a criticism. Market reaction to stalemate depends on whether the markets collectively feel they need something from Washington (stimulus in recent years, perhaps entitlement reform in future years).

If stalemate is a more frequent outcome, then markets will be more likely to deliver bad news than will politicians. This is particularly true on the budget deficit. It's almost inconceivable that politicians will act in advance of market pressures to do so. It's likely to take an acceleration in long-term interest rates or a declining dollar to prompt even an effort to shrink the budget deficit.

To the extent one-party progress is the outcome, then policies are likely to be more volatile. Political scientists have used the "median voter rule" to explain a wide range of policies, which assumes that the legislative process roughly reflects the policy preferences of the voter at the midpoint of the liberal-conservative political spectrum. When policymaking was dominated by centrists in both parties, this model worked well. But in the current polarized environment a "median majority party voter rule" seems more appropriate. That is, when congressional leaders try to craft House and Senate majorities from the majority party, policies will reflect the views of that party. Tax policy is the clearest example, since budget and tax policies have been run through the reconciliation process in recent years.

8

The U.S. Fiscal Deficit—Not a Moment to Lose

WILLIAM C. DUDLEY
EDWARD F. MCKELVEY

WILLIAM DUDLEY AND EDWARD MCKELVEY own the U.S. deficit-analysis franchise. As chief U.S. economist and senior U.S. economist at Goldman Sachs in New York, Dudley and McKelvey's work is given equal weight—often, greater weight—than the best analysis Washington can offer. Out of complex economic models, they deliver a simple message: crisis. Dudley and McKelvey offer well-charted solutions, courses navigable by the individual politician but treacherous to consensus Washington. The ship's bell chimes. Dudley and McKelvey are at the helm of deficit analysis, and they state with force: There is not a moment to lose.

The U.S. budget outlook is grim, despite prospects for near-term cyclical improvement. Over the next ten years, the federal deficit is likely to total $4.8 trillion, or about 2.8 percent of gross domestic product (GDP), double the average since World War II. Thereafter, the outlook gets worse, for two reasons. The cost of making recent tax cuts permanent climbs sharply, and outlays for Social Security and Medicare will surge as the baby-boom generation retires.

Chronic deficits threaten U.S. economic performance. All else equal, they reduce national saving, raise interest rates, and crowd out private investment. The resulting damage to capital formation hurts productivity growth and the potential growth rate of GDP. In turn, slower GDP growth boosts the deficit, completing a vicious circle of self-reinforcing negative dynamics.

Given the seriousness of this problem, all options should be on the table in seeking to resolve it. A reasonable program would include the following steps:

1. Aim to cut the deficit by more than half by the end of the decade;
2. Reinstate the discretionary caps and "pay as you go" rules that helped produce surpluses in the 1990s;
3. Make any further extensions of recent tax cuts contingent on budget improvement;
4. Direct future tax reform toward broadening the tax base and shifting toward consumption-based taxation;
5. Fix Social Security by changing how benefits are calculated and by increasing the retirement age;
6. Put a dent in Medicare by promoting healthier life styles and creating incentives for health care users to be more price sensitive and health care providers to be more efficient; and
7. Pursue growth-enhancing policies such as trade liberalization and investment in human capital.

The Budget Outlook—A Trillion Here, a Trillion There

The federal budget is finally benefiting from strong economic growth following four years of sharp deterioration. The events behind that deterioration of recent years—tax cuts, a surge in spending for national security, and economic weakness—are now "in the numbers." Thus, the fiscal 2005 deficit is likely to come in $60 to $70 billion below last year's record $412 billion.

However, the improvement will be small and short-lived—probably lasting no more than another year or two. According to estimates by the Congressional Budget Office (CBO), only about 15 percent of this year's deficit is due to cyclically low economic activity and other transitory factors. Although the CBO's January 2005 baseline estimates showed the budget moving into balance early in the next decade, they are based on artificial assumptions—current law for entitlements and taxes, including a presumption that recent tax cuts are allowed to expire, and holding constant the real value of current services for programs subject to annual appropriations.

A more realistic outlook would show little or no improvement over the next five to ten years, even with economic assumptions more favorable to the budget than those underlying the CBO baseline. Four other

FIGURE 8.1 **The Long-Term Budget Outlook—No Lasting Improvement**

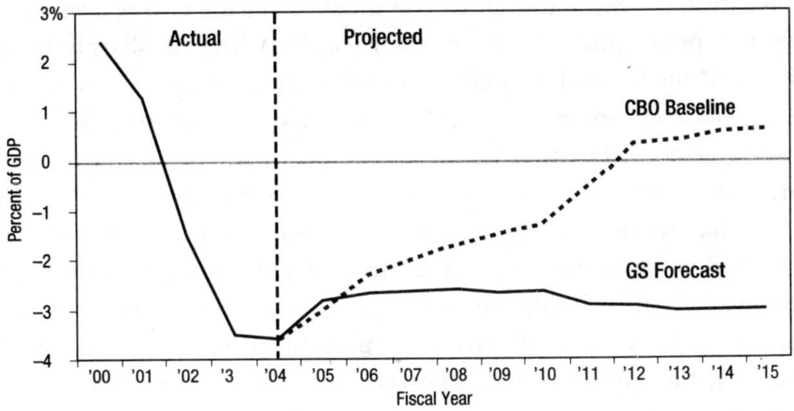

Source: Congressional Budget Office; Goldman Sachs

modifications, designed to transform the baseline into a forecast more in tune with current political realities, point to a cumulative shortfall of about $4.8 trillion over the next ten years, a far cry from the CBO's $980 billion figure. These modifications (1) make the recent personal income tax cuts permanent; (2) index the alternative minimum tax (AMT); (3) reduce defense spending gradually to 3 percent of GDP; and (4) allow other discretionary outlays to grow 2 percent per year in real terms. As shown in ***Figure 8.1***, such a path would keep the deficit between 2.5 percent and 3 percent of GDP. Moreover, risks lie to the upside, as these adjustments are on the conservative side.

Beyond the next ten years, the outlook gets much worse as the retirement of the baby-boom generation causes outlays for Social Security and Medicare to surge. The resulting budget pressures are aggravated by the ongoing increase in life expectancies, the tendency of Medicare spending to rise faster than nominal GDP on a per-capita basis, and the added costs of the new prescription drug benefit enacted in 2003.

For Social Security, the impending retirement of baby boomers and the lengthening of life expectancies are the key factors. Together, they will push the proportion of the population that is 65 years and older to almost 20 percent in 2030, from about 12 percent currently, according to the Trustees' intermediate case assumptions. As a result, spending on Social Security is projected to rise to 6.1 percent of GDP in 2030, from 4.3 percent today. Beyond that, the Social Security spending trajectory gradually levels out. However, the passing of the baby-boom generation does not cure the Social Security problem because life expectancies are projected to continue increasing.

Medicare spending is apt to rise much faster for two reasons. First, the new prescription drug benefit drives spending up sharply beginning in 2006. Second, in addition to the sharp increase in beneficiaries, per-capita costs are projected to continue rising faster than GDP. This "excess cost" differential is due to the peculiar economics of health care, which encourage innovations to improve health care outcomes rather than to control costs. Even on the assumption that this differential slows to an average annual increase of 1 percent (from the 3 percent average since 1970), Medicare spending is projected to rise to 6.8 percent of GDP in 2030, from 2.6 percent currently. Unlike Social Security, the Medicare spending trajectory does not flatten once the baby-boom generation is gone, because the excess cost component continues to push up the share of GDP.

Although public attention focuses on the years when the trust funds for Social Security and Medicare are expected to be exhausted, these entitlement programs will exert pressures on the budget balance much sooner. That is because the impact on the overall balance is determined by the gap between receipts (excluding interest income on trust fund balances) and outlays. Interest income is excluded because it is merely an intragovernmental transfer that has no impact on either the overall budget balance or the government's external financing needs.

By this metric, pressures from these programs are close at hand. For the Social Security Old Age and Supplemental Disability Insurance (OASDI) fund, the gap between receipts net of interest and outlays is apt to peak in 2008. Thereafter, the net positive contribution from this program to the budget balance diminishes. For the Medicare Hospital Insurance (HI) fund, the peak occurred during 2004. By the time these fund balances hit zero—in 2041 for OASDI and 2020 for Medicare HI—the gaps between receipts and outlays will be deeply negative and therefore big drags on the budget balance.

Outlays for Medicare SMI, the supplemental medical insurance part of the program, are projected to grow very rapidly as well, jumping sharply in 2006 when the prescription drug benefit is implemented. Because the SMI programs are mainly funded out of general revenues, their burden on the budget is proportional to the rise in overall expenditures (75 percent in the case of SMI, Part B). Total outlays for SMI are currently projected to climb to 4.2 percent of GDP in 2030, from 1.2 percent in 2005. Of the 4.2 percent in 2030, 1.7 percent is for the prescription drug benefit.

Everett Dirksen, the Illinois Republican who served as Senate minority leader during the 1960s, once said of the budget: "A billion here, a billion there, pretty soon it adds up to real money." In the past forty years, only the units have changed.

If Budget Deficits Don't Matter, Why Do We Pay Taxes?

Some policymakers are unfazed by forecasts of burgeoning federal deficits. Their arguments go along two lines. One is simply that budget forecasts are highly inaccurate and therefore unreliable. For example, in 1992 the deficit was larger relative to the economy than it is today (about 5 percent of GDP then, versus about 3 percent now), and chronic deficits were projected over the next decade. However, by 1998 the budget had moved into surplus. Couldn't this happen again? Couldn't large projected deficits vanish as the U.S. economy continues to grow?

The second claim is more pointed and substantive—namely, that "deficits don't matter." At the moment, proponents of this view appear to have the facts on their side, as evidence that the deficit is hurting U.S. economic performance is scant. Interest rates are low, inflation is low, and the economy is growing at a solid pace.

The answer to these points starts with the simple fact that now is not 1992. The cyclical component of the deficit is smaller now than it was then, and the baby-boom generation is on the verge of retirement. Thus, prospects that economic growth will cause the deficit to evaporate are much dimmer. Also, while long-term budget forecasts are inherently uncertain and subject to potentially large revisions, they are not biased toward showing deficits that are too large. The failure to forecast the emergence of surpluses in the 1990s was followed by an equally spectacular failure to forecast the reemergence of deficits in the current decade.

Put differently, the size of the current deficit is not the key concern. A sharp cyclical deterioration was inevitable as the 1990s investment boom turned into a bust. Fiscal stimulus was warranted to contain economic weakness. Without it, the economy would have weakened even more and the deficit would still have been large.

The problem is this: In implementing this stimulus and meeting other national objectives, the United States has embarked on a set of policies that will make the deficit chronic and damage U.S. economic performance if left unchanged. This damage occurs through the following self-reinforcing process: The deficit represents a claim on national saving that reduces the funds available for private-sector investment. Interest rates rise in the process, to ration the smaller pool of funds among alternative investment projects. If large deficits persist, less is invested in the economy, and the rate of capital deepening slows. In turn, this slows the growth rate of productivity, reducing the potential growth rate of real GDP. Slower growth in GDP means slower growth in tax revenue, which deepens the budget deficit. The result is a vicious circle of high deficits begetting poor economic performance, which in turn begets bigger deficits, a further deterioration in economic performance, and so on.

This view is not as widely accepted as it should be for two reasons:

First, large budget deficits take time to do their damage. Temporary deficits due to cyclical weakness do not cause lasting problems. Because the impact of chronic deficits builds slowly over time, the tie between the deficits and economic weakness is not visible to the naked eye. Moreover, this tie has several links. The validity of some of these links is not fully appreciated.

Second, some view large deficits as a necessary evil to achieve a greater good—namely, to reduce the size of government. According to this view, the political will needed to reduce the government's claims on the economy can be generated only by making the deficit unacceptably large. In this case, so the argument goes, the benefits of reducing these claims will ultimately outstrip the costs imposed by the budget deficits.

This last argument may have some validity. After all, the large deficits of the 1980s ultimately led to the Budget Enforcement Act of 1990, which did help control spending during the 1990s. However, this is scarcely optimal policy. If smaller government is an important goal, it would be better to cut spending directly and run a much smaller budget deficit. The feasibility of this is a political issue, and advocates should press their case in that arena rather than through the back door of large deficits.

As for the other reasons, the circle from large deficits to poor economic performance and back to larger deficits can be established fairly easily. As shown in ***Figure 8.2***, it has seven links. Some of these links are uncontroversial; others have been subject to considerable dispute.

FIGURE 8.2 **A Vicious Circle—Deficits Hurt Growth, Raising Deficits**

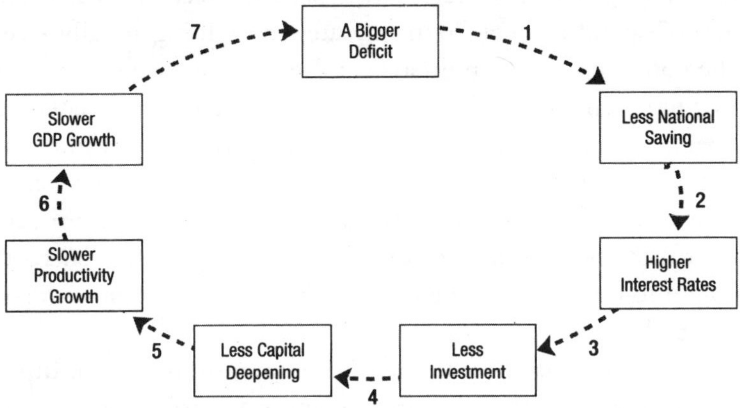

Source: Goldman Sachs

Link #1: Large budget deficits reduce national saving. This link is straightforward because the deficit is a claim on national saving. Unless saving by other domestic sectors moves dollar for dollar in the opposite direction to changes in the federal budget balance, an increase in the deficit will lead to lower national saving. Since 1970, net national saving has fallen about 70 cents on average per extra dollar of deficit.

Link #2: Lower national saving boosts interest rates. This link is actually a red herring in the debate about the deficit's impact on the economy. Once the negative link between the deficit and national saving has been established, the effect on private investment follows through the saving/investment identity. Higher interest rates are merely the pricing mechanism by which the more limited supply of funds is rationed and, in that sense, an obvious consequence of higher deficits. The linkage here is obscured by the fact that the business cycle affects both interest rates and deficits. When one controls for the effects of the economy, higher budget deficits result in higher interest rates.

Link #3: Higher interest rates deter private investment. The business cycle obscures this link, just as it does the link between budget deficits and interest rates. When the economy is growing rapidly, investment spending is high, reflecting pressure on existing capacity. The increase in investment plus other pressures on capital markets usually push interest rates up. Thus, investment does not exhibit a clear negative correlation with the level of interest rates.

However, causality also goes the other way. An increase in interest rates raises the cost of capital and, at the margin, deters some investment. The higher level of rates is the pricing mechanism that rations the pool of capital available to the private sector. It is generally accepted that the cost of capital is an important driver of investment spending.

How much investment falls in response to a rise in the federal deficit depends on the net reduction in aggregate saving—national saving plus foreign capital inflows. In the 2003 *Economic Report of the President,* the Council of Economic Advisers (CEA) posits that for each $1 increase in the budget deficit, total saving would fall 60 cents.[1] Thus, a $100-billion increase in the deficit would lower investment spending by $60 billion.

Link #4: Less investment means less capital deepening. This link is basically a truism, as net investment is, by definition, the change in the capital stock. If gross investment is high enough to cover both depreciation in the existing capital stock and growth in employment, then each worker will have more capital to work with, which is defined as capital deepening. Otherwise, the ratio of capital to labor will fall. Because neither the depreciation rate nor the growth rate of employment depends in any important way on the rate of (current) investment, capital deepening will occur less rapidly when investment weakens in response to higher budget deficits.

Link #5: Less capital deepening leads to slower productivity growth. The link between capital deepening and growth in labor productivity—output per hour worked—should be intuitive. If workers have a bigger and better capital stock to work with, they are likely to produce more per hour. This link was apparent in the composition of productivity growth during the 1990s. In the first half of the decade, capital deepening contributed only 0.5 percentage point per year to productivity growth. By the second half, after investment spending had risen sharply, this contribution more than doubled, to 1.1 percentage points.

Link #6: Slower productivity growth leads to slower economic growth. This link also should be transparent, as potential real GDP growth equals labor-force growth plus the growth rate in productivity. The only way that potential real GDP growth could not move up and down in lockstep with productivity growth would be if labor-force growth moved consistently in an offsetting fashion. However, since 1960, labor-force growth has exhibited a small, positive correlation with lagged growth in productivity.

This completes the chain from large budget deficits to slower economic growth; but is the effect big enough to matter? A simple calculation suggests that the structural deficit, currently put by CBO at about 2.5 percent of GDP, will exert a growth drag of about 0.3 percentage point per year, based on the CEA's estimate that each $1 of deficit lowers private investment by 60 cents.[2]

Three-tenths of a point per year on real GDP growth doesn't sound like much, but compounded over time it makes a huge difference. For example, if such a structural deficit stayed in place for fifty years, the calculation implies that GDP would be roughly 15 percent lower by the end of this period than it would have been with a balanced budget. Moreover, this is not the end of the story, because of the last link, which completes a vicious circle of negative self-reinforcing dynamics.

Link #7: Slower GDP growth means bigger deficits. On the assumption that federal revenue is 20 percent of GDP at the margin, this would cause a revenue shortfall of about 3 percentage points of GDP by the end of the fifty-year period, resulting in a still-higher budget deficit. This implies an even bigger negative impact on capital formation, which implies slower growth and an even bigger budget deficit. Because this incremental deterioration would unfold progressively over the period, the ultimate effects on both GDP and the deficit would be even larger.

Obviously, this vicious circle has the potential to turn a sustainable budget path, defined as one keeping the ratio of federal debt to GDP stable, into an explosive, runaway path featuring a sharp and ultimately unsustainable rise in this ratio and in net interest expense as a percentage of GDP. The United States' heavy dependence on capital inflows from abroad poses a big risk in this regard. Implicitly, the growth effects calculated above assume that capital inflows reduce the deficit's impact on private investment to about 60 cents on the dollar. However, this impact could widen if foreign investors' willingness to hold dollar assets waned, and it could potentially exceed 100 percent if capital inflow turned into capital flight severe enough to overcome any domestic saving offset to higher deficits. The calculation also makes no allowance for increases in federal interest expense, which would occur both because the debt itself is rising and because the growing instability leads to higher interest rates. In short, by damaging domestic economic performance, higher budget deficits generate —via several channels—self-reinforcing, negative dynamics for both federal finances and the economy.

What to Do—Hope for the Best, Plan for the Worst

So what can be done? It is imperative that steps be taken immediately to improve the near-term budget outlook, in order to put the federal government on the best possible financial footing before the baby boomers start to retire. Entitlement reform is also essential for a viable longer-term budget outlook, and the sooner this is done the better as well. Finally, measures designed to boost the nation's growth rate can help resolve the imbalance and should therefore be part of the strategy.

The following seven-point program would go a long way toward achieving these aims:

1. Raise the bar on near-term deficit reduction—Budget deterioration has finally spawned a political consensus to cut the deficit over the next five years. In the 2004 presidential campaign, both President Bush and his Democratic challenger, Senator Kerry, embraced this objective.

Although ambitious from a political standpoint, this is a minimal goal from either a cyclical or secular standpoint. Five years from now, the expansion would be more than eight years old. By that time in the last business cycle, the budget had already been in surplus for a year. From a secular point of view, the urgency of cutting the deficit should be obvious. In 2009, the first baby boomers turn 63—two years shy of the traditional retirement age.

2. Reinstate budget control mechanisms—In 1990, the first Bush administration and the Congress hit on a formula for budget discipline that proved enormously successful. Renewed in 1993 and 1997, the "pay as you go" (PAYGO) system set caps on discretionary outlays and required that legislation proposing tax cuts or spending increases include measures to offset the anticipated budget effect. In essence, PAYGO assured that revenue surprises would flow to the bottom line. However, the discipline unraveled soon after the budget moved into surplus. PAYGO was allowed to lapse in 2002, even as renewed budget deterioration was already underway.

PAYGO should be reinstated immediately. To be effective, it must apply to both sides of the ledger. It should also be a permanent feature of the budget process rather than just a five-year provision that applies for the term of the budget resolution, as was previously the case. This cannot insure against suspension of PAYGO, which at times may even be necessary for cyclical reasons, but it might provide more resistance to departures that do not have strong cyclical justification.

3. Make extension of tax cuts contingent on budget improvement—Although the recent tax cuts were enacted with sunset clauses to contain their budget impact, most observers assume they will be made permanent. This would be extremely costly. Tax cut extension is the most expensive of the adjustments to the ten-year CBO baseline noted earlier, and the cost balloons in subsequent years.

If the budget news turns out to be better than expected, then there may be room to make some of the tax cuts permanent. Alternatively, if the economy disappoints, or if the failure to extend tax cuts threatens to weaken the economy, then short-term extensions or phased sunsets may be justified. However, to commit to making the tax cuts permanent before the relevant budget or economic outcomes are known is premature at best and damaging to the nation's fiscal and economic health at worst. The approach should be "hope for the best, but plan for the worst," as opposed to the current approach of some policymakers, which seems to be "plan for the best and pray you don't get the worst."

4. Reform taxes to stimulate saving, investment, and labor-force participation—Future changes in the tax code should aim to broaden the tax base, keep marginal tax rates low, and move toward a consumption-based tax. As part of any such reform, the current income tax and payroll tax systems should be combined. This would make it easier to change the tax system without dramatically altering its progressivity.

The budget outlook is dire enough to warrant a thorough review of all tax preferences. President Bush has appointed a bipartisan commission to conduct such a review. Since the preferences benefit different groups of taxpayers, the commission's proposals should be considered as a package.

5. Fix Social Security with modest changes to retirement ages and benefit formulas—Any long-term fix for the budget must include changes to Social Security and Medicare—more funding, lower benefits, or a combination of both. Social Security is relatively easy to reform, at least conceptually. A combination of raising the retirement age, trimming benefits available to early retirees, and/or tying the benefit calculation to prices rather than wages could be sufficient to restore solvency indefinitely.

The Bush administration advocates setting up a system of individual retirement accounts funded by a portion of the payroll tax that now finances Social Security, a form of partial privatization. On an actuarial basis, this helps solve the problem because these accounts would be projected to earn higher returns than U.S. Treasuries.

However, privatized accounts are not a free lunch. Higher projected rates of returns reflect the judgment that the assets are riskier. Put differently, there would be a much wider dispersion of potential returns, and actual returns could be lower. Also, if payroll taxes were diverted into these accounts, then some other source of funds would be needed to cover the unfunded liabilities of the current system.

6. Create incentives for Medicare beneficiaries to adopt healthier lifestyles and become more price sensitive, and for health care providers to become more cost conscious. Medicare is a tougher nut to crack for three reasons. First, it is difficult to cut expenditures without reducing the quality of health care for seniors, and political pressures currently push the other way, judging from the recent passage of a prescription drug benefit. Second, the savings from increasing the retirement age are much smaller for Medicare than for Social Security because Medicare outlays for 65- and 66-year-old seniors tend to be much lower than for older seniors. Third, the costs of providing Medicare services are rising faster than overall inflation.

However, options do exist. Coupling private medical savings accounts and catastrophic health insurance coverage is one approach that has merit. In 2003, Medicare legislation took a significant step in this direction. Medicare should move farther toward a managed care system with capitated (that is, fixed cost per enrollee per year) reimbursement. This creates incentives for health care providers to control costs and eliminate services that do not improve outcomes. However, it is important that both the private savings account and managed care approaches be designed so that the reduction in costs results from changes in behavior and improvements in efficiency rather than phantom savings that occur just because the healthiest seniors select the private savings account or managed care options. Finally, changes in malpractice liability are advisable in order to reduce the number of unnecessary medical procedures and tests that do not improve health care outcomes.

Ultimately, some rationing of health care for seniors may prove necessary. For example, queuing would reduce unit costs by helping to facilitate the fuller utilization of health care resources. Another approach would be to establish an overall budget, with the health care services eligible for reimbursement rank-ordered from highest priority to lowest priority. Services would be covered and reimbursed based on ranking until the annual budget was exhausted. (The State of Oregon, for example, has followed this approach in its Medicaid program.) The advantage is that the spending would be prioritized by the efficacy of the health care procedure.

7. Pursue further trade liberalization, human capital investment, and other growth-enhancing policies. In trade policy, the objective should be to minimize the barriers to free trade. Most economists agree that free trade leads to higher levels of output and employment. At a minimum, Congress and the administration should renew their commitment to the Doha round of multilateral trade talks, dismantling of barriers to agricultural trade, and liberalization of trade in services. Prospects of job losses in various industries and perceptions that working conditions in some parts of the world create unfair cost advantages create political obstacles. However, increasing aid for education and retraining is a far better way to deal with these matters than adopting protectionist measures.

Another worthy approach is to increase investment in human capital. If workers have more skills, productivity growth will be higher and the economy bigger. Options include more aid for job retraining or greater investment in children's programs such as Head Start.

Other ideas undoubtedly deserve consideration, but significant advances along these lines would go a long way toward putting federal finances on a much sounder footing. With the baby-boom generation just seven years away from retirement, there is not a moment to lose.

Chapter Notes

1. See 2003 *Economic Report of the President,* Council of Economic Advisers, p. 56.

2. For more details on this calculation, see William C. Dudley and Edward F. McKelvey, "The US Budget Outlook: A Surplus of Deficits," Goldman Sachs Global Paper #106, March 31, 2004. Implicitly, the counterfactual comparison is one in which the federal government would pursue a growth-neutral mix of tax and spending policies to achieve a balanced budget.

9

Corporate Profits—Critical for Business Analysis

RICHARD B. BERNER

RICHARD BERNER crafts economic reports. His average essay is most economists' fondest hope. Berner, chief U.S. economist at Morgan Stanley, writes tight, sequential paragraphs, always keeping aggregate theory grounded in microeconomic foundation. Within the ferment that is the tradition of Morgan Stanley global economic research, Berner is the optimist. He believes in the American experiment, always questioning consensus gloom. That he does so with first-rate intellect, curiosity, and clarity is to our advantage. Here, Berner sheds light on American business and its inextricable linkage to the larger economics of the United States and the world.

As business analysts, many economists produce macroeconomic forecasts that—for all their shortcomings—are key ingredients in business planning and in the analysis of the direction of financial markets and monetary and fiscal policy. Analysts and the business press rightly focus on GDP and its components as the cornerstones of both macroanalysis and measurement. Getting that analysis mostly right is critical for our companies and our careers.[‡]

In my view, however, analysts spend too little time analyzing the behavior of corporate profits and returns on investment. It's hardly their

‡ This chapter was developed from ideas first presented in the author's Presidential Address at the annual meeting of the National Association for Business Economics (NABE) in September 2001. The author is grateful to Robert Crow, Steve Galbraith, Marty Leibowitz, and Henry McVey for helpful suggestions.

fault. The typical undergraduate textbook on macroeconomics mentions profits at most twice—once in outlining the National Income Accounts, and once briefly in discussing investment. Small wonder that some analysts think a focus on profits is important only for economists who work on Wall Street, given its obsession with the stock market. Analyzing profits, however, is also critical for economists who work on Main Street.

There are three key reasons to focus on corporate profits and profit margins. First, and most obvious, the link between Wall Street and Main Street through profitability, stock prices, and their influence on the economy has never been more apparent. Second, analyzing profit margins helps identify what I call critical tension points that may or may not be reflected in market prices. Let's face it: Our efficient capital markets may not mis-price assets for long, but there's no denying the bubble we experienced in the late 1990s. Analyzing these stress indicators often shows what might have to change—in the economy, markets, or in our companies. Last, and most important, current and expected profitability both reflect—and are key drivers of—investment, hiring, and pricing decisions. Thus, far from being just a byproduct of economic forecasts, profits and margins should be an integral determinant of the outcome.

Many economists might think this judgment too harsh. After all, profit maximization is a central tenet of modern macro, as well as micro, theory, yielding important insights about how companies behave. Even before Modigliani and Miller, for example, analysts have recognized that a firm's valuation and its investment policy are inseparable. But conclusions from that neoclassical paradigm often miss the mark for the real world in which we operate. Looking directly at profitability and rates of return can help crosscheck those conclusions. Likewise, returns on investment are key determinants of the strategic recommendations that economists make to their senior management. But too often economists' microanalysis lacks a macro profits context to frame the discussion.

To other economists, schooled in the business cycle analysis of Wesley Mitchell, Arthur Burns, and Geoffrey Moore, or of the Austrian School, a focus on profits is second nature. Corporate profits at one time were thought to be a leading economic indicator. According to Mitchell, that is with good reason; his theory of the business cycle put profits at center stage. A period of economic expansion would boost profits and thus business investment. But as operating rates rise, costs would accelerate faster than output. The resulting squeeze on profit margins would curb investment, triggering recession. Cost cutting in the slump would restore margins and lay the groundwork for recovery. No doubt, many managers would accept this as a reasonable description of today's business cycle.

For many analysts, however, the proof of the pudding is whether profits and returns matter. For the investor, it's obvious: returns are ultimately one of just two things that matter (the other is of course what investors are willing to pay for them). For the macro forecaster, I offer two pieces of evidence that show profitability as a key driver of investment outlays. First, as ***Figure 9.1*** shows, the business cycle has always driven both profit margins and changes in investment outlays. That covariation masked the separate influence of profitability on investment—until recently. The simultaneous boom in profits and capital spending in the 1990s, the more recent bust, and the subsequent recovery all revealed that profit margins do influence investment. But is it mere correlation? Tentative econometric evidence from Macroeconomic Advisers suggests that profitability, in addition to other factors, drives investment. However, the jury is still out on whether it improves forecasts.

FIGURE 9.1 **Profits—A Leading Indicator?**

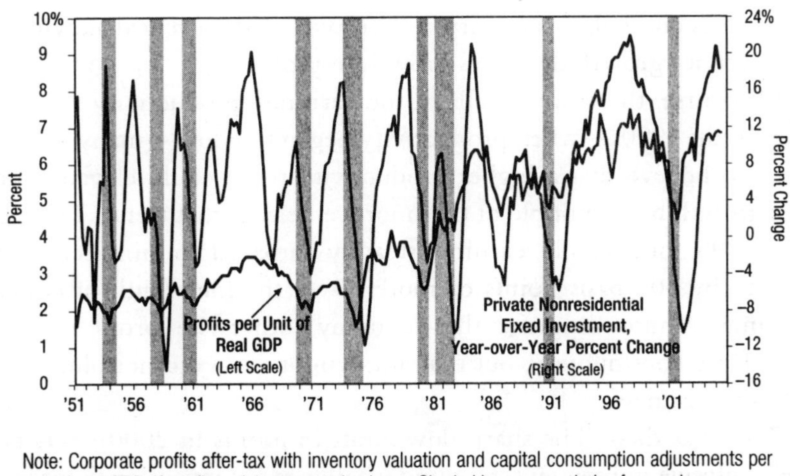

Note: Corporate profits after-tax with inventory valuation and capital consumption adjustments per unit of real GDP of nonfinancial corporate business. Shaded bars are periods of recession.

Source: Bureau of Economic Analysis

I'm not suggesting that the causation from profits to investment is one-way. Far from it. The interplay works in both directions. Indeed, as the discussion that follows makes clear, the relationship going in the other direction—from investment to profits—is complex: Too little investment will starve growth, but too much investment will destroy profitability and returns. But to ignore the profits–economy connection breaks a key link in the analytical chain.

Tools of Profit Analysis

Given the importance of profits, what is the best way to forecast and analyze their connection to economic activity? Most business analysts look to both revenues and costs, figuring that profits are the residual. Because labor costs account for the bulk of the total for many businesses, it's natural to focus on the growth of unit labor costs and its two components, compensation and productivity, as key determinants of earnings growth.

The relationship between earnings and productivity growth is important for analysis. A higher productivity trend is good news for corporate profits. It translates one-for-one into a higher rate of long-term earnings growth, because it means that the economy's potential, or longer-term, sustainable growth rate is higher than before by the amount of the trend productivity improvement (but note that this does not hold for earnings per share, as discussed below). Likewise, a cyclical productivity surge, commonly seen at the start of recovery, will cut labor costs and help boost earnings growth, and a cyclical slowing in productivity growth depresses growth in profits.

Of course, there are pitfalls in the earnings–productivity relationship. In the 1990s—when productivity began to boom—many analysts came to believe that a higher productivity trend could elevate earnings growth by a multiple of the improvement in that trend. Through early 1999, for example, earnings steadily outpaced nominal economic growth by 400 basis points or more. Over the long haul, of course, earnings cannot outgrow the economy, otherwise profit margins would rise continuously, but that annoying truism did not blunt late-1990s bullishness.

That was then. The sharp downdraft in profits in 2000–2001 and the explosive subsequent rebound challenged conventional tools of profits analysis and strongly suggested that the relationship between profits and the economy changed over the past decade or more. *In my*

FIGURE 9.2 **Corporate Financial Leverage—Now Falling**

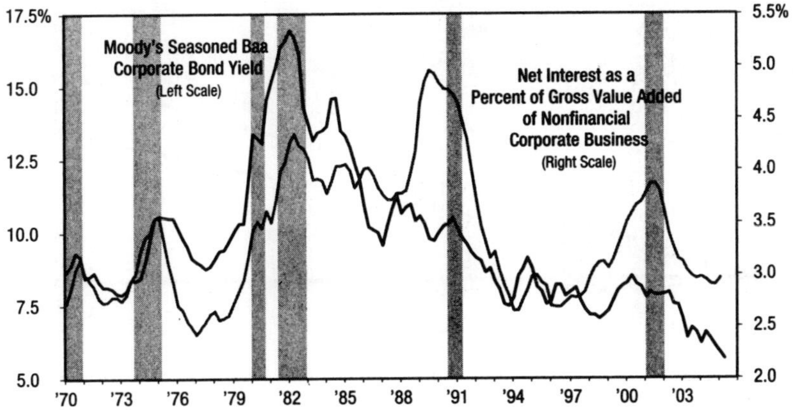

Note: Shaded bars are periods of recession.

Source: Bureau of Economic Analysis; Federal Reserve

opinion, profits have steadily become more sensitive, or leveraged, to growth. Profit margins are a helpful metric for gauging that changing relationship and its connection to returns on investment.

Two factors explain those changes. First, Corporate America significantly increased both financial and operating leverage over much of the 1990s. Second, U.S. profits are increasingly dependent on overseas economic activity.

Let's examine financial and operating leverage over the recent cycle. Levering up boosts returns in expansions, especially if companies buy back stock. But bond investors look askance at financial leverage. That's because highly levered businesses suffer in downturns as debt service eats into margins and increases the risk of default. Eroding credit quality, especially in telecommunications, was a hallmark of the recent recession. Debt service for some companies rose to high levels in relation to cash flow, and investors demanded punitive premiums to take on higher credit risks. Both squeezed profit margins. As ***Figure 9.2*** shows, however, the rise in debt service in the last expansion paled by comparison with the leveraged buyout (LBO) boom of the 1980s, thanks partly to lower interest rates. So financial leverage wasn't the only culprit behind the collapse in margins in the recent recession.

Instead, the real story lies in operating leverage. Finance 101 teaches us that the change in operating leverage is the change in the

FIGURE 9.3 **Cutting Variable Costs and . . .**

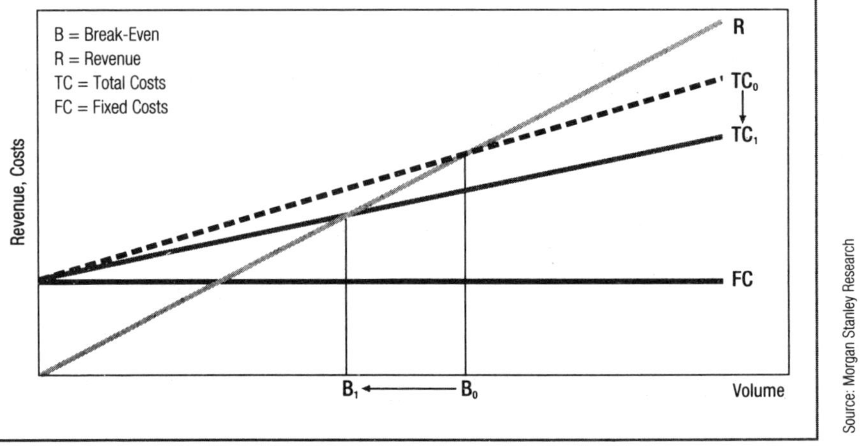

FIGURE 9.4 **Hiking Fixed Costs Boosts Operating Leverage**

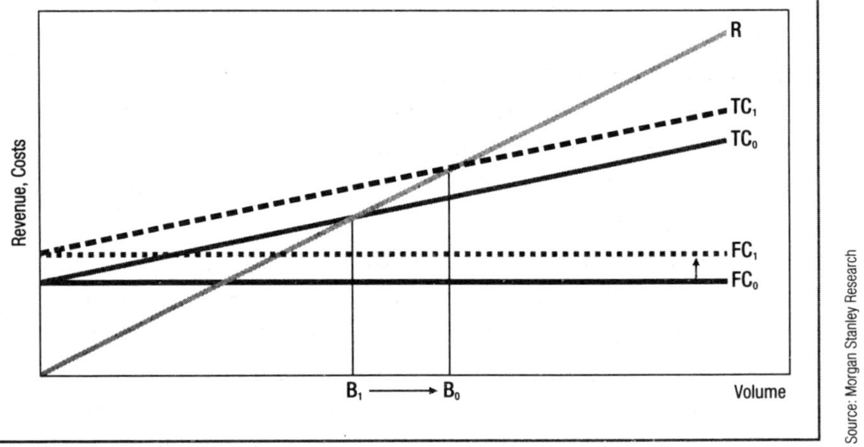

ratio of fixed to total costs. That ratio increased for most companies in the 1990s, as Corporate America scaled back break-even points and boosted ROE by investing heavily in labor-saving equipment. Courtesy of lower marginal, but higher fixed costs, many companies are left with increased operating leverage as the overlooked legacy of this capital spending boom.

In my view, this increased leverage crushed earnings in the 2001 recession by more than had occurred in past slowdowns. Companies must pay the fixed costs regardless, and the boom's most obvious legacy

—extra capacity—translates into increased competitive pressures via plunging operating rates. Symmetrically, the increased operating leverage boosted earnings in recovery by more than in the past. As companies spread those fixed costs over a wider revenue base in 2002–2003, cost per unit plunged, and margins exploded.

Figures 9.3 and ***9.4*** show how operating leverage works. In Figure 9.3, cutting variable costs scales back the break-even point from B_0 to B_1. As evidence of increased leverage, at volumes to the right of B_0, profit per unit is higher than before. Figure 9.4 shows what happens if that reduction in variable costs was achieved through investment that raised fixed costs. The break-even point is back to B_0. But profits are much more sensitive to volumes.

Depreciation—a fixed charge that results from stepped-up investment and that fades slowly—in relation to sales is one metric for gauging the fixed costs associated with increased operating leverage. As ***Figure 9.5*** shows, depreciation as a share of sales at information technology (IT) and materials companies in the S&P 500 was on a rising trend over the 1984-1999 period. For both groupings, that ratio peaked at the recession's trough and plunged in recovery. Likewise, as seen in ***Figure 9.6***, corporate depreciation relative to corporate GDP measured in the National Income and Product Accounts (NIPAs) escalated steadily in the 1990s, and plummeted in recovery.

FIGURE 9.5 **Measuring Leverage—Information Technology and Materials (S&P Sectors)**

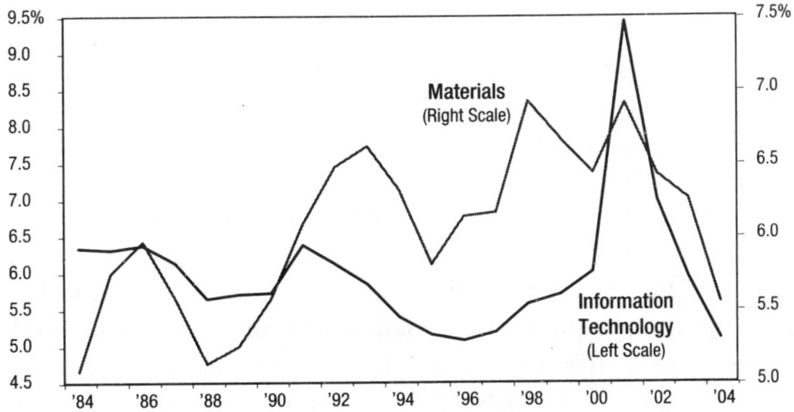

Note: Depreciation as a percentage of sales.

Source: Morgan Stanley Research; Compustat

FIGURE 9.6 **Measuring Leverage—Economy-Wide**

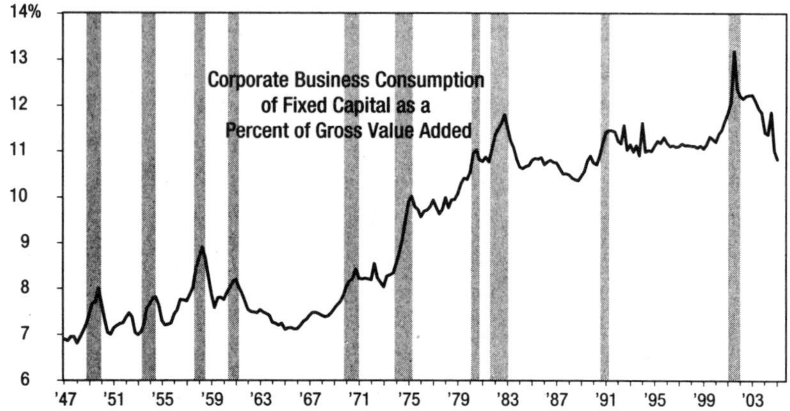

Note: Shaded bars are periods of recession.

Source: Bureau of Economic Analysis

Capacity-Leverage Interplay

The interplay between capacity and operating leverage magnifies the impact of leverage on margins as growth slows, as ***Figure 9.7*** illustrates. That's because low and falling utilization rates also hurt profit margins, not just once, but twice. First, industries with extra capacity can rarely raise prices and are often forced to lower them. Second, low utilization rates are often associated with cyclically inefficient operations and, thus, rising marginal costs as volumes shrink.

The recent gap between economy-wide profit margins and factory operating rates depicted in Figure 9.7 also highlights the fact that America's services businesses have increased their operating leverage in the past few years. Many, such as airlines and other transportation services and telecommunications, have always been capital intensive, thus had high fixed to total costs and high operating leverage. But deregulation and the cost-cutting that ensued—especially via substituting capital for labor—increased operating leverage. And other services businesses, such as wholesale and retail trade and financial services, have increasingly used technology to boost efficiency and inventory turnover and to take advantage of scale. All these businesses thus have increased fixed costs and operating leverage.

FIGURE 9.7 **Capacity-Leverage Interplay Drives Margins**

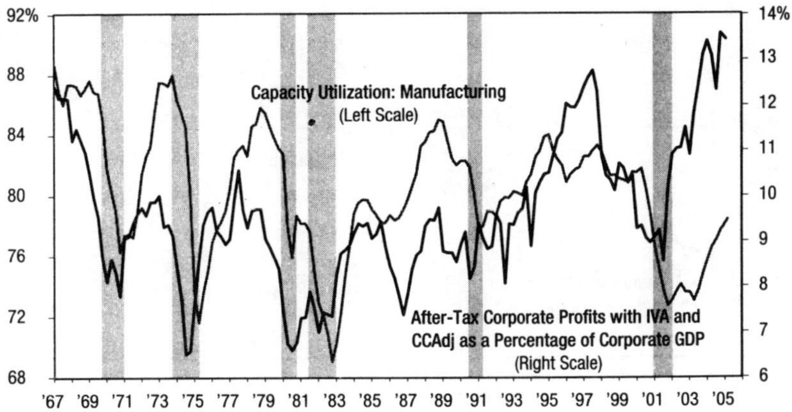

Note: IVA = inventory valuation adjustment; CCAdj = capital consumption adjustment. Shaded bars are periods of recession.

Source: Bureau of Economic Analysis; Federal Reserve

Even in the past, when operating and financial leverage were not such significant issues, sinking utilization rates were bad news for profit margins. In the recent recession, the slide in operating rates was nothing short of stunning. Courtesy of the investment boom, the capacity buildup in some industries—many in technology-producing segments and in communications, as well as in the purchasers of high-tech gear—went to excess. And in other industries, such as primary metals and chemicals, demand weakened by enough to expose significant pockets of excess capacity. The result was the lowest operating rates in manufacturing since 1983, and for technology producers, since 1975.

What many analysts have failed to recognize is that this excess triggered aggressive corporate cutting back of capacity, or "capital exit," across the economy. The combination of the capital spending bust and the passage of time that depreciated installed capacity brought capacity growth to three-decade lows and set the stage for a vigorous rebound in operating rates when recovery took hold. And of course, the capacity-operating leverage process in recovery boosts margins dramatically, for the same reasons it crushed them in recession. Rising operating rates help restore pricing power and efficiency, giving margins a double boost.

FIGURE 9.8 **Stock Returns Are Linked to ROIC**

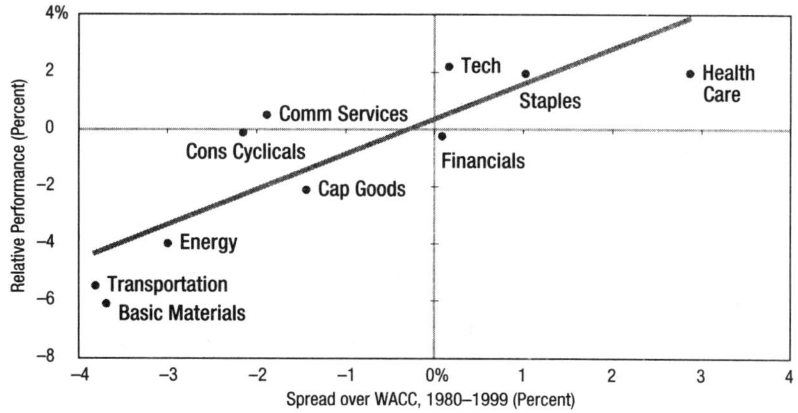

Source: Morgan Stanley Research. Data assembled by Steve Galbraith

This interplay between capacity and leverage has broader implications for macroanalysis, for investors, and for corporate managers. Companies or industries that are disciplined in their use of capital will realize superior returns on assets or equity, while those that are profligate builders or that are serial acquirers will see subpar results. Economic and finance theories teach us that companies that consistently earn returns in excess of their capital costs should outperform for investors. So the market should reward the first group with relatively higher share prices and punish the second group with declining market value. The data in ***Figure 9.8***, assembled by my former colleague Steve Galbraith, bear that out. Industry relative stock market performance is plotted on the vertical axis, while returns on invested capital (ROIC) relative to the weighted average cost of capital (WACC) are plotted on the horizontal axis.

What is really informative about this figure isn't the correlation, but which industries fall in which quadrants. Over the two decades that ended in 1999, industries that used capital carefully—such as health care, consumer staples, and, yes, technology—produced superior returns in both dimensions. The key here is that the companies in the Northeast quadrant sustainably and consistently maintained high ROICs. Meanwhile, those unable to right-size their industries, especially if they were in cyclical, commodity-producing groups, disappointed their managers and investors alike.

FIGURE 9.9 **ROIC, Information Technology**

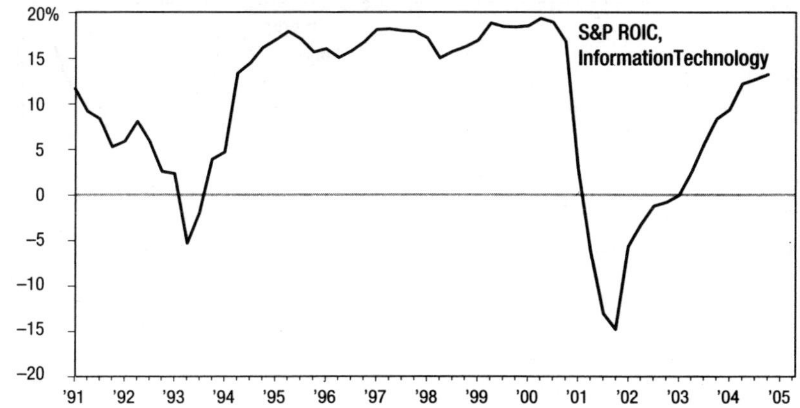

Source: Morgan Stanley Research

That was then. Many technology industries and their customers, perhaps responding to years of blowout returns, became what Steve Galbraith calls capital pigs. As we now know, they overinvested in capacity, ignoring the fundamentals of finance. Had they heeded the first signs of eroding returns in 2000, as depicted in ***Figure 9.9***, many of them might have managed the downturn more smoothly. In comparison, the folks in the energy business have turned from pig into cash cow, as their approach to adding capacity has remained more disciplined than in the past. As seen in ***Figure 9.10*** on the following page, there's no doubt that action taken by OPEC also hurts or helps their performance. But there's no mistaking the fact that, so far, they haven't responded to higher energy prices with a flood of new capacity—and they've sustained higher returns over the past four years.

In short, there are two lessons in this for analysts and managers. One is obvious, but overlooked: Returns and margins matter for analysis and decisions, and may even be leading indicators of performance. The second may be more surprising for the macro analyst, but is second nature to portfolio managers: Good stewards of shareholders' capital are usually good performers, so relative levels of capital spending are good leading indicators of—and often are inversely related to—sector and company performance.

FIGURE 9.10 **ROIC, Energy**

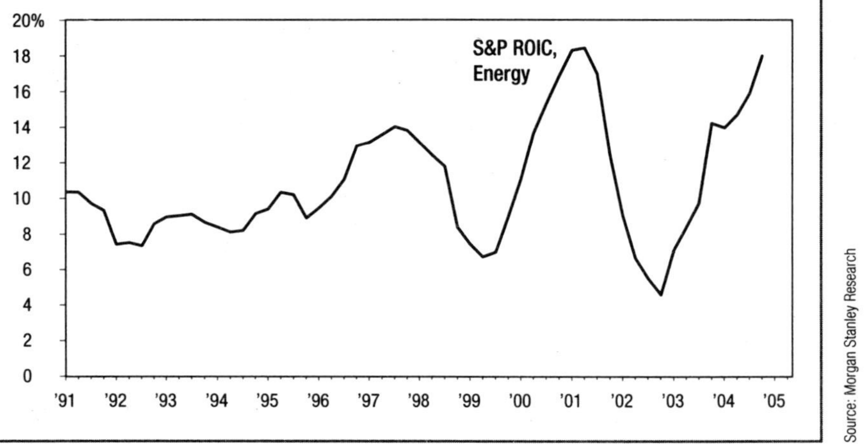

Global Influences

The second factor that has changed the relationship between profits and the economy is the fact that U.S. earnings are increasingly leveraged to global, not just U.S., growth. That should come as no surprise. Many U.S. companies either operate globally or have some suppliers that are located abroad. And exports account for 10.2 percent of GDP today—double the share four decades ago. Less obvious is the fact that over the same time frame U.S. companies' overseas affiliates have also more than doubled their share of overall corporate profits, to 25 percent. As seen in ***Figure 9.11*** at right, that share is clearly cyclical, rising in recessions as domestic corporate income turns down. But it is also rising on a secular basis, as globalization knits our companies and economies more closely. That's a channel from growth to earnings that my former colleague Joe Quinlan has long highlighted with good reason.

Will the Real Corporate Profits Please Stand Up?

Refining our techniques of analysis, and extending the analysis to valuable bodies of information like corporate profits, is critical to success in making forecasts and decisions. But that analysis is obviously no better than the data on which it is based. It is clear that, as good as our statistical infrastructure is in the United States, there is substantial room for im-

FIGURE 9.11 **Rising Share of Earnings from Abroad, 1960–2005**

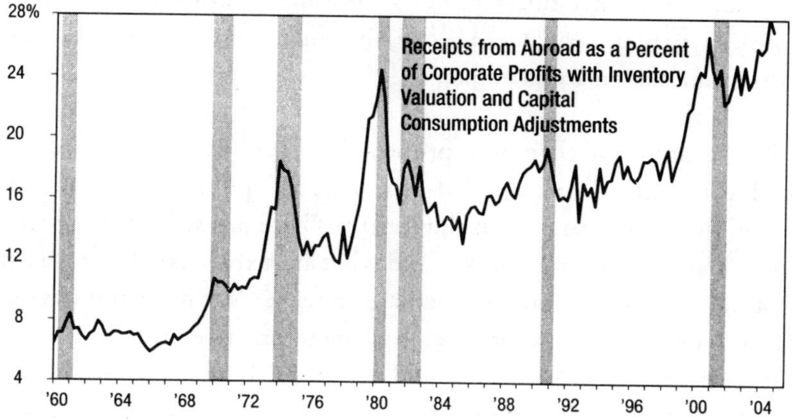

Note: Shaded bars are periods of recession.

Source: Bureau of Economic Analysis

provement. In a landmark address at the National Association for Business Economics (NABE) Washington Economic Policy Conference in March 2001, Federal Reserve Chairman Alan Greenspan said:

> The experience of the last forty years underscores a fundamental dilemma of business economics. Should we endeavor to continue to refine our techniques of deriving maximum information from an existing body of data? Or should we find ways to augment our data library to gain better insight into how our economy is functioning? Obviously, we should do both, but I suspect greater payoffs will come from more data than from more technique. (See "The Challenge of Measuring and Modeling a Dynamic Economy," *Business Economics,* vol. 36, no. 2, April 2001, pp. 5–8.)

And as I noted in congressional testimony as NABE president in April 2001:

> Since their inception, statisticians have endeavored to improve the quality and accuracy of these statistics. Yet our economy is constantly changing: The industrial economy of the past has given way to a very different, knowledge-based information economy. That constant evolution—some would say revolution—requires both new sources of

> data and the resources for our statistical agencies to collect and analyze them. While U.S. economic statistics remain among the best in the world, lack of investment in our statistical infrastructure has left us with a system that still does a better job of measuring industrial activity than information-based output.

In the arena of corporate profits, this rapid change toward idea-based value added has clouded our sense of today's results and thus limited our ability to reckon the future. There are at least three sets of issues: What are revenues today, and what are expenses? What is a capital outlay, and what is an intermediate expense? What is compensation, and how do options and other gain-sharing practices affect it?

The biggest problems surround expenses, both for capital outlays and compensation. Decisions about what to expense and what to capitalize will obviously affect depreciation and earnings. The problem is fuzziest when dealing with intangible assets, like ideas or information, because their cost may bear little relationship to their value. But even with tangible assets, the concept of useful lives in the information technology age bedevils measurement of earnings. How should we capitalize software expenses?

Regarding compensation, few dispute that the accounting for stock options overstates earnings, essentially because their fair value as compensation is not charged against income. Lacking any independent measurement of the value of stock options exercised, U.S. government statisticians were forced to deduce them from reports on wages and salaries. For example, four successive dramatic and offsetting revisions to NIPA corporate profits and compensation during the 1998–2000 period show how this lack of information can significantly distort reality. Partly because they uncovered more option income than previously recognized, statisticians trimmed the level of corporate profits in 2000 by a cumulative 8.9 percent. In response, the statisticians have devised better ways to approximate option grants and exercises. While it is harder to prove, efforts to inflate earnings may well have misled statisticians in Washington when they made their first estimates of corporate profits (for details, see Charles P. Himmelberg, et. al., "Revisions to Corporate Profits: What We Know and When We Knew It," Federal Reserve Bank of New York, March 2004).

These revisions had stunning implications for profit margins. The new data revealed that companies in 1997–2001 endured the most intense margin squeeze since 1978–1980. The old data were bad enough: They showed after-tax "economic" profits as a share of corporate GDP—one proxy for margins—sagging by more than 200 basis points to just under 10 percent between the fall of 1997 and early 2001. Revised data now show that margins plunged by a total of 420 basis points between 1997 and 2001, or roughly one third, to 8.5 percent. That was their lowest level since 1992.

Confusion about accounting for options is only one of several issues that cloud the measurement of earnings in the real world, let alone in the national income accounts. What are recurring and nonrecurring charges against income? In 2001, Standard & Poor's figured that second-quarter operating earnings fell by 32.9 percent from the previous year, while Thomson Financial/First Call put the decline at just 17 percent. How could the gap have been so wide? A few tech companies took sizable write-offs in the second quarter of 2001; Standard & Poor's counted these as an operating expense, and First Call saw them as one-off items.

More ominously, those who carefully plied the accounting trade in 2000–2001 began to unearth several ways that companies overstated earnings. Morgan Stanley head of Global Valuation and Accounting, Trevor Harris, who was at the forefront of this forensic work, noted five at the time:

1. They represented one-time sales as ongoing revenue;
2. They recognized revenue before goods were shipped;
3. They assisted customers with vendor financing but did not account for its cost;
4. They used questionable assumptions in accounting for pension costs in operating income; and
5. They overstated restructuring charges following an acquisition.

The confusion over earnings, and the suspicion that they had been overstated for years, couldn't have come at a worse time. Amid the gloom over a slumping economy and a hard landing for corporate profits, investors sensed in the summer of 2001 that stocks may have been even more overvalued than they previously thought. The surfacing of corporate scandals and suspected fraud further undermined investor confidence. Indeed, the S&P 500 plunged another 20 percent over the next two years. Trevor Harris' work showed that companies that play games with earnings won't be able to hide results. It should now be clear that they weren't just hurting their shareholders. They hurt themselves and the broad investing public.

What should be done? I applaud the Bush administration and Congress for appropriating increased funding for our statistical agencies to improve the quality of economic statistics. Still more is needed to try to solve some of the biggest shortcomings of our statistics. Business can partner with government to help. NABE is initiating efforts to enable businesses to contribute to the solution. Let's also make sure that businesses don't contribute to the problem. We should insist on transparency for our policymakers, as they are now beginning to insist on transparency in corporate reporting.

Earnings in a World of Single-Digit Returns

I believe that we are in a world of single-digit returns across most asset classes. Seeing the secular or longer-term glass as half full may seem ironic in view of my debate with Steve Roach over the U.S. economic outlook. But that debate concerns the near-term cyclical outlook; when Steve and I look ahead to longer-term secular developments, we have much in common (see "Common Ground," *Global Economic Forum,* May 25, 2002). That we are now in such a world is rooted in my belief that economic reality will put a tether on long-term investment returns. This has important micro and macro implications.

The bear market in equities ended in 2003, but investors' expectations for long-term investment returns may still be too high. Likewise, investors are probably still too optimistic about medium-term earnings growth. Echoing the 1990s' confusion over the relationship between earnings and productivity, many seem to think that lasting economic recovery implies that earnings gains will return to a sustainable double-digit pace—that is, for several years to come.

It's worth repeating the laws of arithmetic: Earnings even in a growth-oriented universe like the S&P 500 cannot outstrip the economy's sustainable nominal pace unless profit margins expand—and margins cannot expand without limit. In turn, with medium-term productivity gains likely to be in a 2.5 percent to 3 percent range, a growing labor force, and roughly 2 to 3 percent inflation imply a sustainable 5.5 percent to 6.5 percent range for nominal GDP growth.

Make no mistake, the productivity surprise of 2002–2003, coupled with high operating leverage, promoted the margin explosion during those two years, as in the second half of the 1990s. However, margins over the business cycle will probably be roughly stable now that they have more than recovered the cyclical ground lost in recession. If that is the case, nominal GDP and earnings growth are unlikely to diverge significantly. Indeed, in the past decade, so-called "economic" profits after-tax rose by 8.2 percent, while nominal GDP gained 5.3 percent. Clearly, the profits bust in the recession wiped out much of the outperformance of the late 1990s.[1]

In my view, we've come full circle to a "back-to-basics" stock market. I believe in reversion to the mean, and that it will take time for sustainable earnings gains to grow into valuations. The implications of such an environment are profound: Dividends and yields matter again, and fixed-income securities deserve significant weightings in balanced portfolios. This not-so-brave new world may trigger changes in compensation and pension policies and require rethinking public policy for retirement saving—all of which has implications for earnings and returns.

For equity investors, dividends may be especially important. The interplay between the double taxation of dividends and spectacular capital gains induced Corporate America to slash dividend payouts over the past decade. That explains why the S&P 500 dividend payout ratio hit a record low 31.8 percent in 2000. Modern finance theory dictates that apart from tax issues, investors should be indifferent between capital gains and dividends because managers could at least equal market returns through reinvestment or they could distribute returns through share buybacks. But, quite apart from the allure of dividends following the reduction in dividend taxes, investors lacking confidence in management may now seek companies that boost dividends because they want the right to decide how to reinvest earnings. And in the post–Enron world, dividends may signal a commitment to shareholders that has recently been lacking. As Steve Galbraith was fond of saying, "Share buybacks are like dating, dividends are like marriage."

In that sense, Microsoft's decision to pay out a one-time special $3 per share dividend, boost its regular dividend, and end its stock option program was a watershed. This is not just a high-tech growth company suddenly deciding that, free of litigation risk, it lacks investment opportunities. It is a company that wants to be known as a good steward of shareholder cash and a superior allocator of capital.

A second implication of single-digit returns is that employee compensation policies may change in a single-digit return world. In a nutshell, employees may grow fonder of cash up front. In the 1990s, a new flexibility emerged in employee compensation. In lieu of higher basic pay, both employers and employees eagerly embraced "gainsharing" that gave a portion of a company's returns to workers. Thus employees shared in the fruits of corporate performance in good times, and employers could spread the pain of recession to total compensation—perhaps with fewer layoffs—when recession hit. The tax and accounting treatment of stock options encouraged their issue for a broader range of employees. But with many options granted in the bubble years still under water and the Financial Accounting Standards Board mandating changes in accounting for options, their popularity seems to be waning. It's reasonable to ask whether employees will want either restricted stock or more compensation now, and in cash.

Third, single-digit returns have important implications for corporate pension policies. Throughout the 1990s, a rising market led to income from defined benefit (DB) pension plans for many old-line companies. And the same trend led to overfunding of those plans on an actuarial basis. So big was the cushion of overfunding and so onerous the penalties for making contributions to "overfunded" plans that companies took funding holidays. That and overoptimistic return assumptions are the two most insidious pension legacies of the 1990s. In 2001, Corporate America assumed that pension plan assets would return 9.2 percent over the long run. Although it is single-digit, that assumed long-term return is probably too high. If so, pension fund income is likely to provide less support for future corporate profits; indeed, contributions may absorb corporate resources. Recent return shortfalls, lower returns, and underfunding have forced companies to step up plan contributions, eroding profits. Strapped companies are threatening to break pension promises. The bankruptcy of several steel companies and airlines whose legacy costs for retirees proved fatal is an extreme example, but the tension between benefits and resources is likely to grow.

There's little doubt in my mind that the analysis of corporate profits and the appropriate measurement of earnings are issues that are important for both Wall Street and Main Street. Compared with past business cycles, secular—as opposed to traditional cyclical—themes are legitimately a more important part of the economic and investment landscape. No doubt, a single-digit return environment will reshape thinking about many other investment norms, including the dollar's value in foreign exchange markets. This new trend may also reshape our thinking about, and have consequences for, the macroeconomic environment we face in the next decade. Equally, however, it is important not to lose sight of the cyclical forces—including changes in profitability—that drive the economy's evolution. Both are important.

CHAPTER NOTES

1. Even if that outperformance had continued, per-share earnings growth would probably decline. EPS is determined by the reinvestment rate of the firm, *not* the rate of output growth. Output growth cannot translate permanently one-for-one into higher per-share earnings growth because firms must raise capital through issuing more shares or debt to fund the investment in new technology that partly fuels the improvement in productivity and, thus, sustainable output growth.

10

Monetary Policy, Wicksell, and Gold

JOHN RYDING

JOHN RYDING is the outlier. Where others balance the certitude of various and sundry Keynesian pasts with a most-certain Laffer present, Ryding chooses not to participate. He is the market economist most associated with the optimism of a visible-market solution. Observe, and believe in markets. That is the Hayekian, Ryding remedy. As chief U.S. economist for Bear Stearns, Ryding is critical of Federal Reserve thinking and the prescriptive wave that cascades down upon Chairman Alan Greenspan—you should do this; you should not do that. Here is Ryding's neo-Wicksellian essay on the folly of disbelief in market information. Ignore the markets—particularly gold—at your peril.

> "If...the leading banks of the world were to lower their rate of interest, say 1 percent, below its ordinary level, and keep it so for some years, then the prices of all commodities would rise and rise and rise without any limit whatever."
>
> —KNUT WICKSELL, 1906

> "The evidence indicates clearly that [an accommodative] policy stance will not be compatible indefinitely with price stability and sustainable growth; the real federal funds rate will eventually need to rise to a more neutral level."
>
> —ALAN GREENSPAN, FEBRUARY 2004

What determines the inflation rate is the central issue for monetary policy. In my judgment, the Federal Reserve Board relies too heavily on the model of inflation known as the Phillips Curve, or output-gap. This model views the inflation rate as a function of the gap between actual employment and full employment (a.k.a. the natural rate of employment or the nonaccelerating inflation rate of unemployment or NAIRU).

As a result of this model, when the unemployment rate was falling in the late 1990s—eventually hitting a low of 3.8 percent in April 2000—the Fed was worried about inflation. For example, in May 2000, the Fed's policy statement said, "The Committee is concerned that this disparity in the growth of demand and potential supply will continue, which could foster inflationary imbalances that would undermine the economy's outstanding performance." In May 2000, the core consumer price index (CPI) inflation rate was 2.4 percent, and it increased to only 2.8 percent a year later, before plunging to 1.1 percent by the end of 2003. Meanwhile, the economy fell into recession in March 2001. The Fed's policy statement, however, didn't begin to highlight potential deflation concerns until May 2003, when the statement said, "the probability of an unwelcome substantial fall in inflation, though minor, exceeds that of a pickup in inflation from its already low level." At that point, the core CPI inflation rate was 1.5 percent and would fall less than a half percentage point further over the next six months before beginning to increase.

Fed Chairman Alan Greenspan has a tremendous knack for identifying when things are not going according to the models and then altering the course of monetary policy accordingly. The Fed is to be praised for the large and rapid reduction in interest rates in 2001. However, Alan Greenspan will not always be the chairman of the Federal Reserve (indeed, he must step down from that position in January 2006 unless Congress changes the Federal Reserve Act). Unfortunately, there is no such term-limit on the tenure of the Phillips Curve at the Fed. Greenspan has mixed in his pragmatism (and perhaps some of his lingering beliefs from his days as a follower of Ayn Rand) with the Phillips Curve, which enabled Fed policy to go where the Phillips Curve alone would not have taken it.

Defunct Economists

"Practical men, who believe themselves to be quite exempt from intellectual influences, are usually the slave of some defunct economist. Madmen in authority, who hear voices in the air, are distilling their frenzy from some academic scribbler of a few years back."

—JOHN MAYNARD KEYNES, 1936

Unfortunately, these days Keynes is the defunct economist, and the ideas of classical, or pre-Keynesian, economists are generally underappreciated. Nonetheless, I think classical monetary theorists provide invaluable insights into the outlook for growth, inflation, the optimal path for monetary policy, and the consequences of monetary error. Standing tallest among classical monetary theorists is the Swedish economist Knut Wicksell (1851–1926) who advanced the notion of a natural rate of interest. Wicksell's cumulative process ought to have been placed at the center of modern macroeconomics. Instead, macroeconomics placed the Keynesian liquidity-preference (LP) at the center of interest-rate determination and incorporated it into the standard IS-LM model (investment-saving/liquidity-money). The IS-LM model, however, was only a model that determined real GDP (and thus employment). The determination of the price level or inflation rate came later when orthodox macroeconomics tacked on the Phillips Curve.

This may seem abstract and somewhat dry stuff, but this Keynesian-neoclassical synthesis model is essentially the Federal Reserve Board's model of the U.S. economy that drives the Fed's forecast. In the model, interest rates determine investment and the level of GDP (the IS curve). Monetary policy determines the level of short-term interest rates (the LM curve). Unemployment is determined by GDP, which in turn determines the inflation rate (the Phillips Curve). A full discussion of the FRB/U.S. model can be found in the January 1999 *Federal Reserve Bulletin* article "Aggregate Disturbances, Monetary Policy, and the Macroeconomy: The FRB/US Perspective." The authors of that article note, "In the model, inflation is predicted to decline as long as labor and capital are underutilized and to rise whenever resource utilization is above average."

The thought process that underpins the FRB/U.S. model is also the reason why the Fed kept interest rates too high in the late 1990s, thereby producing the recession of 2001. It is also the reason why the

Fed kept monetary policy too easy in 2003 and 2004. In July 1996, Greenspan, in his semiannual testimony on monetary policy, explicitly pointed his finger at labor market tightness as a source of a potential inflation problem. He said, "Historically, current levels of slack, measured in terms of either the unemployment rate or capacity utilization, have often been associated with a gradual strengthening of price and wage pressures...there are early indications that this episode of favorable inflation developments, especially with regard to labor markets, may be drawing to a close."

When Greenspan gave that testimony, the unemployment rate had averaged 5.5 percent in the first half of 1996, and the core inflation rate measured by personal consumption expenditure (PCE) prices had averaged 1.9 percent over the previous twelve months. Over the next five years, the unemployment rate averaged 4.5 percent and the core PCE inflation rate averaged 1.6 percent. However, even as late as November 2000, the Federal Open Market Committee (FOMC) policy statement was concerned about inflation, saying, "The utilization of the pool of available workers remains at an unusually high level, and the increase in energy prices, though having limited effect on core measures of prices to date, still harbors the possibility of raising inflation expectations...the risks continue to be weighted mainly toward conditions that may generate heightened inflation pressures in the foreseeable future."

As inflation continued to fall into 2002 and the unemployment rate continued to move higher, the FOMC became concerned about potential deflationary pressures. In November 2002—one year into the recovery—the Fed cut the funds rate 50 basis points, to 1.25 percent. Again in June 2003, the Fed cut the funds rate a further 25 basis points, to 1 percent, saying, "the probability, though minor, of an unwelcome substantial fall in inflation exceeds that of a pickup in inflation from its already low level. On balance, the Committee believes that the latter concern is likely to predominate for the foreseeable future." The consequence of this Phillips Curve thinking is that U.S. monetary policy became cranked up to an ultra-accommodative setting, yet the majority of forecasters in March 2004 did not see the Fed raising rates until 2005.

Failure of the Phillips Curve—The Seventies Stagflation

The spectacularly poor inflation forecasting performance of the Phillips Curve should come as no surprise to those who remember the 1970s. During that period, the unemployment rate rose from 3.5 percent in 1969 to a peak of 8.5 percent in 1975 and averaged 6.2 percent over the decade. Core PCE price inflation rose from 4.7 percent in 1969 to a peak of 9.9 percent in 1975 and averaged 5.9 percent during the decade. Thus stagflation—the combination of economic stagnation and inflation—was born. Those of us who studied economics in the late 1970s, found the economic textbooks on inflation hopelessly outdated, and the demise of the Phillips Curve and the Keynesian orthodoxy seemed imminent.

The surprise, therefore, is that the output-gap/Phillips Curve model has re-established itself as economic orthodoxy. The failure of the model in the 1970s is explained as a rise in the natural rate of unemployment (NAIRU), in the same way as the Ptolemaists explained away observations of the motion of the planets that were inconsistent with the view that the Earth was at the center of the solar system by the construct of epicycles.

In a theoretical sense, it can be argued that there is a flaw in viewing the determination of the price level by using models that appear to be microeconomic supply-demand oriented. It is the case that the price of a good in an individual market is determined by the balance of supply and demand. If a good is in excess demand, its price rises. However, the price that is being determined is the *relative* price (say, the price of apples versus the price of oranges). The determination of the overall price level, however, is the relative price of all goods and services versus money. The overall price level is thus inherently a different form of price than a price in an individual market and, conceptually, ought to be determined by the balance between the supply of goods versus the supply of money. Where is money in a Phillips Curve or output-gap inflation story?

The Monetarist Counterrevolution

The failure of the Phillips Curve in the 1970s helped fuel Milton Friedman's monetarist counterrevolution. Friedman's model of inflation was essentially a simple one: *Inflation is the result of too much money chasing too few goods.* This simple catchphrase captures the essence of the inflationary process. However, when this is translated into a model in its simple form, it fails to explain the inflation process. Such a model begins with the exchange equation:

$$MV = PY \qquad \text{(Equation 1)}$$

Where M is the quantity of money, V is the velocity of circulation (the number of times money changes hands in a given period), P is the average price level, and Y is the number of transactions or volume of goods produced (PY can be thought of as nominal GDP and Y as real GDP).

In the Friedman world, real GDP (Y) is determined by real factors only (this is referred to as the classical dichotomy). The velocity of money (V) is stable (or at least predictable), and hence variations in money (M) determine movements in the price level. If we take logs of equation 1 and differentiate with respect to time, we get:

$$m + v = p + y \qquad \text{(Equation 2)}$$

The lower cases indicate percentage growth rates of the corresponding variable. Rearranging equation 2 and assuming velocity is stable (and hence $v = 0$), gives:

$$p = m - y \qquad \text{(Equation 3)}$$

Here the inflation rate equals the growth of money less the growth of potential GDP.

Equation 3 was turned into a policy experiment in both the United States and the United Kingdom during the period 1979–1982. Both countries adopted monetary targeting and, according to the theory, all that was required was to restrain the growth of money and the inflation rate would fall. Unfortunately, the world was not that simple and both countries were plunged into deep recessions, thus demonstrating that the classical dichotomy (that money did not influence the real economy) does not hold in the short run. In the United States, real GDP fell 2.3 percent from the first quarter of 1981 to the third quarter of 1982, while real GDP in the United Kingdom fell 4.8 percent from the fourth quarter of 1979 to the first quarter of 1981.

While inflation is a monetary phenomenon, this strict form of quantitative monetarism has not been a useful model. Central banks, such as the Federal Reserve, reverted back to interest rate targeting in the 1980s and money supply was de-emphasized. Indeed, the Fed has abandoned providing ranges for monetary aggregates such as M2. In doing so, money seems to have been thrown out of the inflation story.

A Wicksellian Synthesis

Wicksell's cumulative process ought to be at the center of models of the inflation determination process. This view of the world postulates that there is a neutral or natural rate of interest that would be consistent at any point in time with price stability or a stable inflation rate. In this model, inflation rises or falls according to whether the actual short-term interest rate is below or above the natural rate. However, in the world in which Wicksell lived, monetary policy was run on a more or less strict gold standard and, therefore, interest rates were determined by the market. Monetary authorities had little discretion over influencing interest rates, which tended to move with the economic cycle. Wicksell's central proposition was a purely theoretical one in the sense that he could not conceive of a significant departure of interest rates from their natural level. Indeed, in the United States, the Fed did not yet exist when Wicksell was formulating his theories.

Wicksell's cumulative process, however, is highly relevant in the modern world of fiat money (money that has no backing in the form of gold or other precious commodity). In today's world the Fed and other central banks set the level of short-term interest rates and there is no guarantee that that rate is the natural rate or even close to it. Indeed, in 2003 and the first half of 2004, the Fed deliberately chose to hold the fed funds rate well below the natural or neutral rate. Furthermore, the Fed's exit strategy from the 1 percent fed funds rate environment has been a deliberately gradualist one and, therefore, one which ensured that monetary accommodation would be maintained for some while.

When combined with a somewhat Austrian school view of the business cycle, which focuses on the central role of capital spending in the business cycle (Joseph Schumpter's famous gales of creative destruction of capital), the Wicksellian approach provides a very useful model of the economy and inflation. The most insightful parts of Keynes' General Theory, which focus on the coordination failure between savers and investors, can be integrated into this framework.

What does a sketch of this model look like? First, movements in capital spending are the principal determinants of the shape of the economic cycle. The short-run dynamics of the economy have Keynesian characteristics (such as consumer spending or inventory behavior). However, the origins of disequilibria are found in Wicksell's process. In this model, it is useful to view the equilibrium real interest rate (or real natural rate) as a function of the marginal productivity of capital. The natural rate of nominal interest can be viewed as the natural real rate of interest plus the expected rate of inflation.[1] Thus we can start with the equation:

$$i^* = f(\rho) + p \qquad \text{(Equation 4)}$$

Here ρ is the marginal productivity of capital, and i^* is the natural rate of interest.

However, the monetary authorities set the short-term nominal interest rate i according to some discretionary process, and the actual rate of real interest, r, is defined as:

$$r = i - p \qquad \text{(Equation 5)}$$

Unless the Fed picks i such that $i - p = r^*$, then the inflation rate will not be stable. In particular, if $r > r^*$, then inflation will fall, and if $r < r^*$, inflation will rise. It immediately follows from this that the neutral nominal interest rate is not a fixed number, even if the factors determining the neutral real rate are unchanged. If the Fed keeps the fed funds rate below its neutral level, then inflation will rise and the neutral nominal interest rate will rise with it. If the Fed kept the fed funds rate above its neutral level, then inflation would tend to fall and the neutral nominal rate would also decline. This means that a constant fed funds rate does not represent an unchanged policy stance unless policy was already neutral. If policy was accommodative ($r < r^*$), then inflation would tend to rise and the neutral nominal rate would also tend to rise. If the Fed left nominal interest rates unchanged, the real short-term interest rate (r) would fall and policy would become even more accommodative.

Identifying Neutrality

The theory sounds simple, and it is; however, putting it into practice is a little more complex since no one knows where the neutral rate of interest is. However, it seems reasonable to suppose that leading indicators of inflation would be rising when interest rates are below their neutral level and falling when rates are above neutral. This is where we circle back to gold and other commodity prices. In Wicksell's day, the price of gold was fixed and short-term interest rates floated. These days, it is short-term interest rates that are fixed and it is the price of gold that floats.

The Recession of 2001—A Wicksellian Perspective

The economy in the late 1990s was driven by strong growth in investment spending, especially on technology equipment. Accompanying the strong growth in investment spending was a rising growth rate of productivity. The average growth rate of nonfarm productivity rose from 1.7 percent during the period 1990–1995 to 2.6 percent during 1995–2000. Income-based measures of productivity show an even stronger pickup as nonfinancial corporate productivity growth rose from 1.6 percent to 3.2 percent on the same comparison.

The economy grew robustly despite the emergence of two headwinds. The first was fiscal, as tax payments rose as a share of GDP. From 1992 to 2000, the share of personal tax payments in GDP rose from 9.6 percent to 12.6 percent. The rise no doubt reflected the passage of President Clinton's tax hikes in 1993, the emergence of strong real income growth generating "bracket creep," and the strength of the equity market generating capital gains and other wealth-related tax payments (such as IRA or 401(k) withdrawals). The second headwind was the Fed's monetary policy actions from 1997 onward. The Fed, concerned about the inflation consequences of a falling unemployment rate, raised the fed funds rate target to 5.5 percent in March 1997 from 5.25 percent.

At the time, gold prices stood at around $350 an ounce (see ***Figure 10.1***). In the wake of the Fed's rate hike, gold prices fell to around $325 an ounce by July 1997. This monetary stringency was too much for many Asian economies whose currencies had been pegged to the dollar and, beginning with Thailand in July 1997, a number of Asian economies devalued their currencies against the dollar, producing a wave of imported price deflation. While the financial market consequences of

FIGURE 10.1 Price of Gold

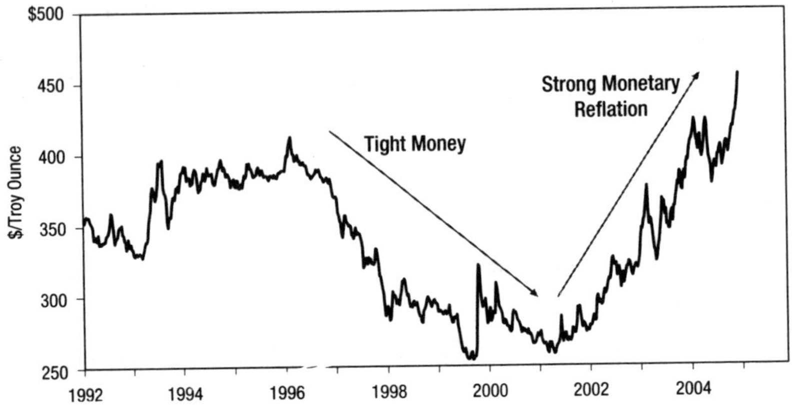

Source: *Wall Street Journal*

these overseas developments stayed the Fed's hand from further interest rate hikes, they did not encourage the Fed to lower rates, and gold prices fell further. By September 1998, gold prices had fallen to around $285 an ounce, clearly showing that the Fed was holding the funds rate above the natural rate. The Fed was forced to cut rates in that month, however, as the LTCM crisis reared its ugly head, and by the end of the year the fed funds rate was at 4.75 percent and gold prices had stabilized.

Once the crisis had passed, the Fed began raising rates commencing in June 1999. With only a short pause because of Y2K-related concerns, the Fed raised the funds rate from 4.75 percent to 6.5 percent by June 2000. Gold prices fell to $255 an ounce in July 1999 and the average price of gold in 2000 was $279 an ounce, which was almost identical to its average price in 1999.

In the summer of 2000, the technology sector of the economy began to falter. Investment in business equipment and software slowed from 15.3 percent in real terms in the first half of 2000 to 0.3 percent in the second half. Manufacturing output fell 4 percent in the second half of 2000, having grown at a 4.5 percent rate in the first half of the year. However, until December 2000, the Fed indicated that it believed the balance of risks was tilted in the direction of higher inflation rather than weaker economic growth.

The Fed began to lower rates in January 2001 as the manufacturing ISM index fell to levels traditionally associated with recession. However, as capital spending began to plunge, with business equipment spending falling at a 10.1 percent rate in the first half of the year, one has to presume that the natural rate of interest also fell. For a while, the Fed's rate cuts could not get ahead of the falling natural rate of interest, and gold prices failed to rise in response to the Fed's rate cuts. It took the tragic events of September 11, 2001, to get the Fed to lower the funds rate to a level that began to stabilize capital spending and reflate gold prices. The 1.75 percent funds rate target reached in December 2001 pushed gold prices up to an average level of $290 an ounce in the first quarter of 2002 and to $313 an ounce in the second quarter. Business equipment spending stabilized in the wake of monetary reflation, which paved the way for the recovery.

The Recovery of 2002–2004—Misplaced Deflation Concerns

The recovery of 2002, however, was unusually sluggish. Real GDP grew by only 2.3 percent in 2002, which, combined with strong productivity growth, resulted in a further loss of about 550,000 payroll jobs in the year. Core inflation also continued to decline in 2002 and 2003, with core PCE price inflation dropping from 2.2 percent in 2001 to 1.7 percent in 2002 and to 1.1 percent in 2003. Payroll employment fell modestly further in 2003, dropping a further 300,000 in the first eight months of 2003.

In response to falling employment and inflation, the Fed cut rates in November 2002 and again in June 2003, taking the funds rate down to 1 percent. Gold prices, however, began to rise further, to $334 an ounce by the end of 2002 and to $408 an ounce by the end of 2003. Commodity prices also recovered during this period, and the value of the dollar in terms of foreign currencies fell sharply.

All of these indicators strongly suggested that the fed funds rate target at this point was below the natural rate, which should eventually lead to higher inflation. The inflation pressures have begun to materialize in 2004 as core CPI price inflation averaged 2.2 percent. At the same time, payroll employment growth has begun to emerge more robustly as payrolls expanded by 2.2 million in 2004.

If the Fed had used a Wicksellian perspective, it would have seen the recovery in gold prices in 2002 and 2003 as a signal that monetary policy was adequately positioned to support the recovery, that policy was becoming easier, and that deflation concerns were misplaced. The

Fed could and should have raised rates modestly in 2003 rather than cutting them further. This would likely have avoided the bond bubble of May-June 2003 and its subsequent bursting in July and August. If the Fed had snugged rates higher in 2003, it is doubtful that we would have seen the same pick up in inflation in 2004.

Going Forward

The Fed is still very much stuck in the Phillips Curve/output-gap framework. The policy statements and speeches of FOMC members refer to slack in labor and product markets and the current low level of inflation as reasons to withdraw monetary accommodation only slowly. However, with the funds rate at 3 percent and nominal GDP growth at 6.3 percent over the last year, the gap between the fed funds target and the natural rate of interest remains very wide (see ***Figure 10.2***). Hence gold prices moved above $440 an ounce—a sixteen-year high—in December 2004.

However, the Fed has continued to push the theme of only gradually higher interest rates. In my judgment, the Fed has a wide gulf to cross before it returns policy to neutral. It seems somewhat inevitable, therefore, that inflation is likely to rise over the next year, eventually pushing the Fed off the gradualist path. If only the policymakers at the Fed would pay more heed to Wicksellian concepts and the message of gold, commodities, and the dollar.

FIGURE 10.2 **Nominal GDP Growth versus Fed Funds**

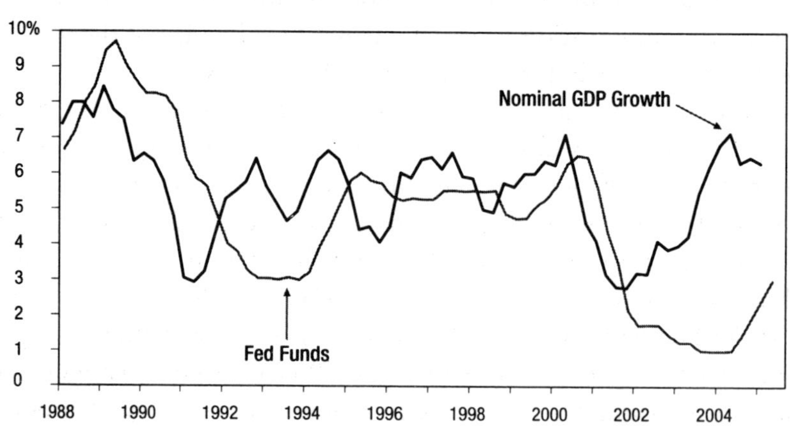

Source: Bureau of Economic Analysis and the Federal Reserve Board

CHAPTER NOTES

1. This model is complicated in a world of taxation. If r is the real rate of interest, i is the nominal rate of interest, p is the expected inflation rate, t is the marginal tax rate and ra is the after-tax real interest rate, then in a world without inflation indexation of capital income

$$ra = i(1-t) - p$$

which can be rewritten as:

$$i = ra/(1-t) + p/(1-t)$$

In other words, if investors are taxed on the inflation component of nominal interest rates and they seek to maintain a constant after-tax real rate of interest, the nominal rate of interest must rise faster than the inflation rate. For example, if the marginal tax rate was 50 percent, then $1/(1-t) = 2$, which means a one percentage point rise in inflation should result in a two percentage point rise in nominal interest rates. It is an empirical question whether interest rates behave in this manner.

11

Commodity Price Pass-Through—Finished Before It Began?

DAVID A. ROSENBERG

DAVID ROSENBERG has an opinion. To legions of Merrill Lynch colleagues and clients, the chief North American economist's terse, day-after-day analysis is agree-or-disagree essential reading. Rosenberg gets away with forceful belief and suggestion, where so many others fail, because underneath his work is a firm set of quantitative skills. Rosenberg links data to economics to markets like no one else. A Rosenberg strength is his analysis of price change—inflation, disinflation, and deflation. Here is David Rosenberg on commodity inflation. Consensus is, once again, catching up to Rosenberg.

The surprisingly high early-2004 core CPI readings seem to have overshadowed the bigger picture trend of what still seems to be a sustained low-inflation environment. What is truly amazing is that we are coming off a two-and-a-half-year period where commodity prices soared—not to mention the sugar-high from unprecedented monetary and fiscal policy stimulus—and yet the pass-through along the supply chain and through to the consumer were the most benign in modern history. To be sure, mid-2004 consumer price reports did show some belated commodity-induced increases, such as higher airfares and perhaps some lagged currency effects on clothing. And it could well be that the typically lagging official inflation statistics drift mildly higher through 2004 a number of companies (Caterpillar, Morton's, Kimberly-Clark, and P&G) announced price hikes in early 2004 to protect margins. But they did so at a time when the commodity cycle looks to be heading into a very mature phase, so a sustained run-up in inflation pressures is unlikely. Copper prices, as of mid-2004, were well off their highs, and global shipping rates had fallen roughly 40 percent. Moreover, the principal source of the commodity inflation originated in China, whose central bank began liquidity-draining initiatives to stem runaway demand growth in early 2004. The other great source of "reflation" in the past two years—the weakening U.S. dollar—began reversing course in early 2004 as well.

The question lingers: Are companies at the final stages of production in the position to offset these higher input prices with higher final goods prices on a sustained basis? (Remember: inflation is a process of repeated price increases, not just once-and-for-all profit margin protection.) Early 2004 inflation reports indicated that companies have never had as much trouble passing on higher costs as they've had in the current cycle. The large amount of slack in the labor market as of mid-2004 (labor force participation rate at a fifteen-year low; employment-to-population ratio at a ten-year low) and product market (sub-75 percent capacity utilization in manufacturing) are hardly the launching pads for a material pickup in overall inflation. And as this chapter will show, the biggest gains in the commodity bull market are behind us, not ahead of us.

In 2004, market attention is increasingly being dominated by the wiggles in the Producer Price Index (PPI) and Consumer Price Index (CPI) monthly data releases. Yet historians are likely to look back at this period and say that it was the first time in more than four decades that underlying inflation hit a cyclical peak of 2.0 percent or less.

Similarities Among All Six Commodity Bull Cycles

The average commodity price bull market lasts for thirty months (see *Figure 11.1*). Guess what? At June 2004, the recent bull market was twenty-nine months old. In terms of average price run-up, the Commodity Research Bureau (CRB) Index historically has risen by 55 percent, though the median is closer to 35 percent because the "mean" is skewed by the massive spike in the early 1970s (reflecting the OPEC, Mideast war, and dollar crises). This commodity cycle, which began in November 2001, has seen the CRB rise by 45 percent. So again, even in terms of magnitude, the best of the rally is probably behind us, not ahead of us. In other words, we are in the very mature stages of this

FIGURE 11.1 **Bull Markets in the Commodity Sector**

Date	CRB Index % Change	CRB Index Annual Rate	Duration (Months)
Nov 71–Aug 74	127.5%	34.8%	33
Oct 76–Feb 80	49.7%	12.9%	40
Dec 82–May 84	28.6%	19.4%	17
Aug 86–July 88	34.2%	16.6%	23
June 93–Aug 96	34.7%	9.9%	38
AVERAGE	55.0%	18.7%	30
Nov 01–April 04	44.3%	15.5%	29

Source: Commodity Research Bureau; Merrill Lynch

bull cycle, and China's tightening of monetary policy may well be the factor that keeps this cycle from becoming much lengthier than its predecessors.

A Uniquely Tame Cycle

What is unique about this cycle is that the pass-through across the supply chain and through to the consumer has never been as tame as is the case currently. Historically, more than 40 percent of the price run-up at the CRB level is ultimately passed on to the PPI intermediate stage and 30 percent to the finished goods level of the PPI. This time around, only 18 percent has been passed on to the intermediate goods PPI and 13 percent to the finished goods level. In other words, this commodity cycle has seen less than half of the typical price response through the producer price pipeline.

As for the overall CPI, it is "normal" to see close to 30 percent of the price rise at the back end of the production curve find its way to the retailer phase—but not in this cycle. What we have seen so far is only 13 percent of the initial price increase being passed on at the CPI level, and most of that is energy and food related. Less than 10 percent of the commodity price increase has been finding its way into the core CPI, which is highly unusual. In the past, the pass-through was 27 percent. So at the core CPI stage, the ability to pass on higher raw material costs is only roughly one-third as strong today as it was, on average, in the prior cycles (see *Figures 11.2* to *11.5*).

FIGURE 11.2 **PPI, Finished Goods**

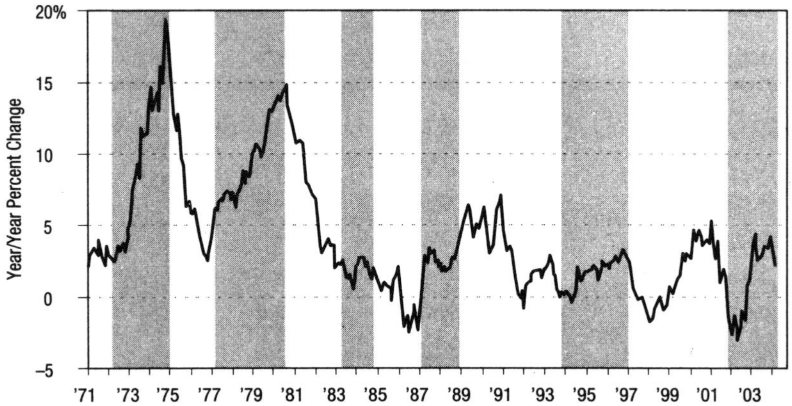

Note: Shaded periods reflect bull markets in commodity prices.

Source: Commodity Research Bureau, Bureau of Labor Statistics

FIGURE 11.3 **PPI, Finished Goods, (ex Food and Energy)**

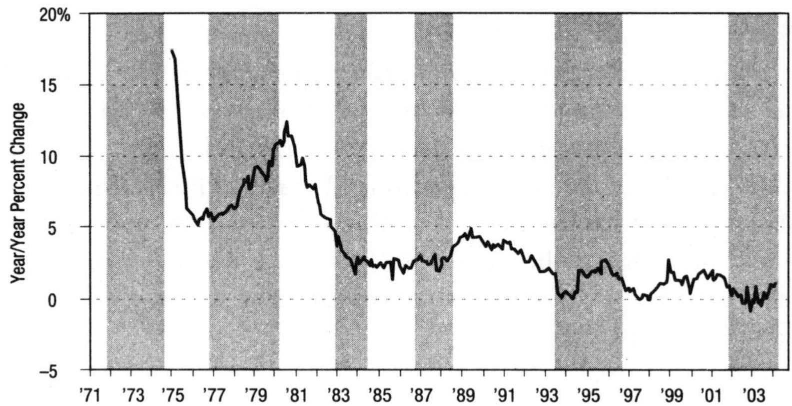

Note: Shaded periods reflect bull markets in commodity prices.

Source: Commodity Research Bureau, Bureau of Labor Statistics

No Exception Among Sectors

Another major difference between this cycle and prior cycles is that this time around the pass-through from higher commodity prices to final goods prices is lower in every sector of the economy. Here are four examples:

FIGURE 11.4 CPI

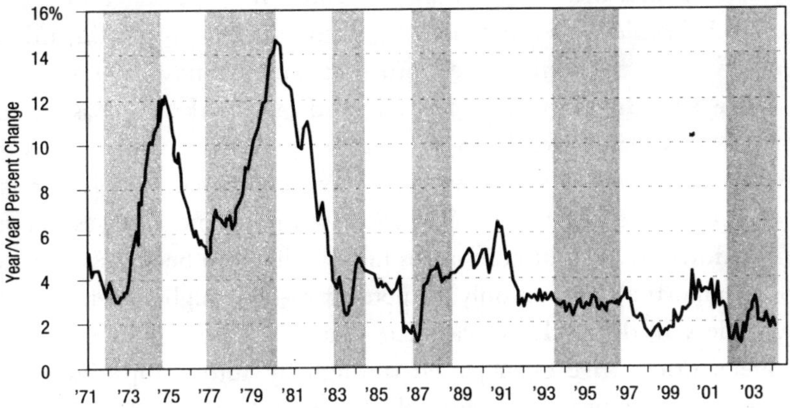

Note: Shaded periods reflect bull markets in commodity prices.

Source: Commodity Research Bureau, Bureau of Labor Statistics

FIGURE 11.5 CORE CPI

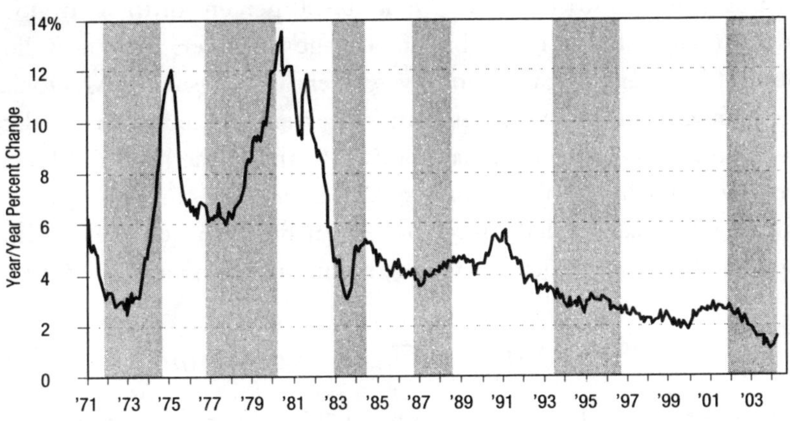

Note: Shaded periods reflect bull markets in commodity prices.

Source: Commodity Research Bureau, Bureau of Labor Statistics

Airlines—The airlines have tended to pass 31 percent of higher fuel costs onto consumer fares. The pass-through this time around has been exceptionally weak. Crude oil prices have risen 87 percent, but airfares have remained roughly flat since November 2001.

Textiles—Usually an average of 11 percent of higher fiber costs gets passed through into the CPI Apparel Index, but this time around, due to the effects of globalization, clothing prices are down 5 percent even though the CRB Textiles Index has risen by almost 20 percent. Again, a divergence we have not seen before—one that is now almost three years old.

Base metals—This has been a record bull market for base metals. The CRB Metals Index is up 86 percent, more than twice the historical norm. The records show that steel/metal companies have historically been able to pass 72 percent of the increase in base metal prices to companies downstream. Car companies historically have been less fortunate. On average, they pass on only 35 percent of their higher metal costs to the dealers. In this cycle, we calculate that only 8 percent of the run-up in metal prices have been passed through by auto companies—about one quarter the usual pass-through. Meanwhile, automakers have had absolutely no luck passing along higher input prices. At the dealer level (read: CPI), new vehicle prices have actually deflated 3.4 percent since November 2001.

Food—The CRB Foodstuffs Index has risen 44 percent during this commodity cycle, 10 percentage points firmer than the average run-up for this group. Historically, we see about 30 percent of the commodity price run-up filtering through to final stage producers (such as Kellogg, General Mills, and Nabisco) and for retailers as well (such as McDonald's and Safeway). In this cycle, less than 20 percent of the raw material cost has found its way into the final goods PPI, and at the retail (CPI) stage, grocery stores have enjoyed only a 12 percent pass-through—less than half of what had tended to get passed on during prior cycles—and a moderately better 14 percent for the restaurant segment of the CPI.

"It's Different This Time Around"

Economists are loath to use the phrase "it's different this time," but in this case—it is. There are two very important differences between the current commodity cycle and prior ones. This is the first commodity boom that coincided with such huge excesses in both the product and labor markets.

In prior commodity cycles, conditions in both markets were materially tighter. As a result of higher commodity prices, workers obtained larger wage gains. Businesses obliged because either job market conditions were too tight or labor unions were too strong—as a result, unit labor costs tended to rise, especially since productivity was rising at a very anemic pace. Companies responded to these wage increases with large and repeated price hikes. This created a vicious cycle.

Today, these conditions simply do not exist. Sure, commodities are in tight supply, but unlike prior cycles, labor isn't. The total unemployment rate—which adds back discouraged workers, marginally attached workers, and those working part time because they couldn't find full-time work—was sitting at 9.6 percent in April 2004, virtually unchanged from April 2003, despite phenomenal growth rates in domestic demand. Even as companies started to hire in mid-2004, wages were barely keeping pace with prices, and as a result real average weekly earnings were basically flat from mid-2003 to mid-2004. There was simply too much slack in the labor market for employers to feel enough pressure to hike wages, and the labor bargaining calendar showed that union/management negotiations were centered more around job security and health care issues than wages and salaries. That is a break from the 1970s experience with runaway inflation, when strong union power and tight labor markets resulted in unit labor costs soaring at over a 7 percent average annual rate (see ***Figure 11.6***).

Moreover, fixed long-term contracts, which tend to guarantee wage increases and cost of living adjustments (a hallmark of the 1970s) have become less prevalent. The unionized share of the workforce is now barely more than 10 percent, a post–World War II low and half the level it was during those "bad" inflation years of the 1970s and early 1980s. The net result is that employers have been able to keep their

FIGURE 11.6 **Unit Labor Costs During Commodity Bull Markets**

	Unit Labor Costs	
Date	**% Change**	**Annual Rate**
Nov 71–Aug 74	21.2%	6.7%
Oct 76–Feb 80	30.2%	7.8%
Dec 82–May 84	0.1%	0.0%
Aug 86–July 88	7.3%	3.6%
June 93–Aug 96	2.5%	0.8%
AVERAGE	**12.3%**	**3.8%**
Nov 01–April 04	–4.6%	–1.7%

Source: Bureau of Economic Analysis and Merrill Lynch

labor costs under control and, as a result, unit labor costs have declined at a 1.7 percent annual rate since the CRB bottomed almost three years ago. This trend has helped keep inflation muted across the pipeline this cycle and is the primary reason why both the Fed and the bond market have been able to "look through" the inflationary implications of the commodity price bulge.

On the goods side, in prior commodity booms industry capacity utilization (CAPU) rates averaged 81 percent. In this cycle, operating rates have averaged around 75 percent, making companies less able to pass on cost increases than they were in the past.

Those trends are not likely to change in the near term. Productivity remains strong, and absorbing the excess slack in the labor and product markets will take a long while.

What If the Commodity Cycle Is Closer to the End Than the Beginning?

There are a few not-so-subtle shifts in the wind which indicate that commodity prices are peaking. Four things look to have changed during the first half of 2004.

First, the Chinese authorities are moving to restrain demand growth. They put through an outright interest rate increase in April 2004, and in May they moved to curb growth by boosting the banking sector reserve requirements to 7.5 percent from 7 percent. Considering that China is now the world's largest consumer of copper, the second largest consumer of aluminum, and the third largest consumer of nickel, any slowing in the Chinese economy could well affect the base metals group.

Second, the trade-weighted U.S. dollar appears to have reached its bottom. From its February 2004 low it actually rallied almost 7 percent by May 2004, as opposed to breaking down the way it did repeatedly through most of 2003, which added impetus to the global reflation theme.

Third, the overwhelming consensus seems to be that we are in the midst of an aggressive synchronized global expansion. But the Organization for Economic Cooperation and Development (OECD) leading indicator slowed from March through May 2004, and the six-month annualized trend at May 2004 receded from the peak rates of growth seen in late 2003. What this metric is telling us is that global production growth will most likely decelerate through the end of 2004.

Fourth, as of May 2004 the Baltic Dry Index (the index of global shipping rates that usually acts as a confirmation indicator for the commodity complex) was nearly 40 percent lower than its previous peak.

The investment focus for so long has been on which sectors to be long and short during commodity bull markets, but the real out-of-consensus call may now be to start focusing on relative sector performance once the commodity cycle hits its peak.

The 2004 CPI data have indeed been a big surprise, and if repeated, would call into question the view that the looming Fed tightening will be truncated. It may well be the case that underlying inflation drifts modestly higher from mid-2004, as the year-over-year monthly base comparisons remained challenging for the rest of 2004. But the key question is whether possible one-off price increases will be sustained. While there are some signs of higher raw material pass-through, inflation is likely to remain benign into 2005.

The bottom line is that we have already seen the greatest easing in monetary and fiscal policy in modern history, a two-year slide in the dollar, and a typical run-up in commodities. And here's what we're left with: As of April 2004, core import prices were at +2.4 percent year over year; core PPI was barely above 1.5 percent, year over year; core PCE (personal consumption expenditure) deflator was at +1.4 percent year over year; GDP deflator was at +1.7 percent year over year in the first quarter; the nonfinancial corporate price deflator was at +0.5 percent year over year in the fourth quarter of 2003; and finally, core CPI in April 2004 was at +1.8 percent year over year. Many pundits complain about the treatment of housing in the CPI given the "imputed rent" method and the large weighting it has in the index. But if you adjust for that by taking the median of the 192 goods and services in the index, this measure is still up 1.9 percent year over year (as of April 2004), while the core CPI is only up 1.2 percent. The upshot is that every major inflation barometer is running within the Fed's definition of price stability—in a 1 percent to 2 percent range—and most are still at the low end of the range. True, the Fed funds rate in May 2004 was at a low 1 percent versus a Taylor Rule estimate of "neutral" that is around 2.5 percent, but it is churning out broad money growth of 4.5 percent, which is perfectly consistent with price stability. If inflation is a case of "too much money chasing too few goods," then we'd ask the big inflationists out there to show us the money!

12

Capital Markets and the Economy

David P. Goldman

David Goldman provides our most challenging chapter. Like most professionals in the capital markets, he assumes basic bond knowledge. For many, this will be heavy lifting. The good news is that Goldman, head of debt research at Banc of America Securities, writes with such verve and velocity that his message of past and future debt/economic "shocks" hits home—even for the mathematically impaired. He approaches the economics of the day from a different angle. Raise the bar, read Goldman. Then try Homer & Leibowitz's Inside the Yield Book: The Classic That Created the Science of Bond Analysis.

Capital market innovations during the past two decades improved market efficiency, reduced the cost of capital, and contributed to economic growth. This coin has a flip side, however. Capital market inefficiencies have led to significant market disruptions, with deleterious consequences for economic growth. This chapter calls attention to several examples of how capital market behavior affects the broader economy.

Capital markets are an expression of the human life cycle: Old people lend money to young people. At the global level, countries with aging populations lend money to countries with young populations, except when the risk associated with young populations discourages such lending. Milton Friedman's permanent-income hypothesis of 1957, the basis of his 1976 Nobel Prize in Economics, provided the first crack in the edifice of the Keynesian school: If individuals adjust their spending according to life-cycle needs rather than momentary changes in their income, Keynesian policy tools would not have the predicted effect. By putting long-term expectations at the center of economics, Friedman's work placed capital markets at the center of attention. Monetary economics as well as the "rational expectations" hypothesis of Robert Lucas both have their origin in Friedman's 1957 challenge to Keynes.

Another challenge to Keynes, namely supply-side economics, proceeded directly from consideration of the role of capital markets in the economy. In 1965, Robert Mundell observed that future household income flows were more uncertain and therefore more difficult to discount in present markets than corporate income flows. For that reason, he argued, an increase in the outstanding volume of government debt might contribute to economic efficiency, because government debt discounts into present markets the future tax receipts from households. Such an increase in efficiency would occur, Mundell argued, in the event that a reduction in tax rates contributed to higher economic growth. Overall tax receipts would decline, but if the increase in tax receipts due to higher growth exceeded the interest cost of the additional debt issued by the government to cover the revenue shortfall, the result would be an increase in efficiency and economic welfare.

Mundell's observation provides a sort of Rosetta Stone between the abstract world of economics and the workaday world of capital markets. Rather than assuming fixed expectations and efficient markets, Mundell's hypothesis assumes that capital markets are imperfectly efficient. The interesting question then becomes: What innovations or policy changes would make capital markets more efficient or less efficient? That has enormous repercussions for economic growth.

Innovations That Improved Capital Market Efficiency

The absence of a banking system in many developing countries, for example, leaves savers with little choice but to bury gold coins in their gardens or the equivalent. If a village has many old people and few young people, the coins will stay buried; if a neighboring village has many young people, they will languish for lack of capital. Something like this characterized the state of the American banking system in the early 1980s, when the rapidly expanding, youthful population of the Sunbelt developed a huge appetite for capital, while the aging savers of the northern Rustbelt lacked local opportunities for investment. Two great innovations matched the income requirements of aging savers with the capital requirements of young families in different regions. They were, of course, the invention of the mortgage-backed securities (MBS) market and the development of interstate banking. Prior to the advent of the MBS market, 14,000 thrift institutions funded local mortgages with local deposits. The thrifts had limited access to sources of capital outside their own local deposit base, for example, by issuing certificates of deposit nationally, but the lending system ultimately depended upon local market knowledge of the lenders and local funding.

Securitization of mortgages solved several problems at once. By pooling thousands of mortgages into a pass-through security under the aegis of the two government-sponsored enterprises, the Federal National Mortgage Corporation and the Federal Home Loan Mortgage Corporation, the MBS market allowed mortgage investors to spread their risk across a geographically diverse portfolio. Second, thrifts with excess loan demand could write mortgages and sell them into a liquid market, for purchase by thrifts with excess deposits. Third, and most important, the existence of a liquid MBS market drew in new classes of investors, including overseas lenders. Not only the excess deposits of the Rustbelt or Florida retirees were available but also (for example) the resources of Japanese banks, who bought a substantial portion of the floating-rate tranches of collateralized mortgage obligations during the late 1980s and early 1990s.

Interstate branch banking, meanwhile, allowed regions with excess savings (such as Florida) to fund loans in regions with rising capital requirements, such as Atlanta. Hugh McColl Jr. of National Bank of North Carolina (later Nationsbank and Bank of America) hailed from Bennettsville, South Carolina, where his family owned the town's sole bank until the Great Depression. No capital flowed through the capillary system from the great national banks or even the regional banks of the South down to towns like Bennettsville. Local economies cannot do without branch banking. Although the MBS market could diversify away the idiosyncratic risk associated with individual borrowers or regions, business loans are a far more heterogeneous entity. Local lending expertise is required to drive business lending. McColl and others broke down the barriers to interstate branch banking during the 1980s and 1990s, allowing deposits to flow to where they were needed.

With Innovations, New Problems

Along with the creation of the high-yield bond market, mortgage-backed securities and interstate branch banking made up the trio of debt capital market innovations that underpinned the great U.S. economic expansion of 1982–2000. Every solution, however, generates its own set of problems. Mobilizing capital from new sources through liquid markets stimulates economic growth. But it also makes the economy dependent on the efficiency of capital markets. Legal, regulatory, and structural issues create fragilities in capital markets, which in turn affect economic performance.

Four characteristics of capital markets deserve attention in the broader economic context:

1. **The optionality of mortgage-backed securities**—Almost alone in the world, American mortgages may be prepaid, or called, at the borrowers' whim. Sharp movements in interest rates can destabilize the MBS market.
2. **Regulatory restrictions on credit quality**—Pension funds and life insurance companies must maintain the vast majority of their fixed-income holdings in investment-grade instruments. Unexpected credit migration toward speculative-grade ratings can create perverse effects.
3. **The migration of the pricing standard for fixed income away from the Treasury curve to the swaps curve**—Volatility in the spread between swaps and Treasuries can translate into credit market instability.
4. **Index-based investing**—When investors gauge their returns against an aggregate benchmark index in fixed income, the results are quite different than in the equities market. The broad equity indices reflect the market's valuation of expected future income streams, because equity prices adjust continuously to market expectations. The composition of the fixed-income market has more to do with regulation, tax effects, and custom than the expected composition of output. Index-based investing can set at odds the economic risk of bond portfolios and the business risk of the investors who manage these portfolios.

The MBS Market

On two occasions in the past decade, sharp swings in the duration of MBS portfolios led to undesirable economic consequences. The first occurred in 1994, when several leading hedge funds (most prominently Granite Capital) collapsed in the wake of Federal Reserve tightening. The second occurred in 1998, when a decline in the value of MBS contributed to the failure of Long-Term Capital Management (LTCM), producing a temporary panic in capital markets.

As financial market participants well know, American law gives homeowners the right to call their mortgages at will without penalty. When interest rates decline, homeowners prepay their mortgages and take out new loans at lower interest rates. Conversely, when interest rates rise, the rate of prepayment slows and the duration of MBS extends. Like Schrödinger's Cat, who dwelt in a superposed condition of being alive and dead, mortgages live in a superposed condition of being a bond and a checking account. There is a small "natural" market for bonds paying extra yield in exchange for an embedded short call position in an option on interest rates, but it is quite limited. Traditional purchasers of bonds, such as pension funds and life insurance companies, require well-defined cash flows in order to meet a well-defined stream of liabilities, namely future pension or insurance payouts.

Mortgage pass-through securities do not satisfy the requirements of traditional bondholders. The financial industry offered a partial remedy to this problem starting in 1984, with the invention of collateralized mortgage obligations, or CMOs. CMOs split a pool of mortgages into tranches offering less duration volatility and more duration volatility, respectively. Up to a given threshold, for example, all prepayments will be assigned to a more volatile "companion" tranche, whose purpose is to reduce prepayments to a less volatile planned amortization class (PAC) tranche. The industry sold PACs to insurance companies and pension funds, and sold the companions to whoever would take on additional prepayment risk.

With the advent of complex models to predict the change in the duration and average life of MBS under a variety of interest-rate scenarios (option-adjusted spread modeling), the financial industry found a wider market for the asset class. Thrift institutions, insurance companies, and other investors employed these systems to create MBS portfolios to match liabilities. The models, the investors believed, would help control average life and duration volatility of MBS portfolios that offered more yield than the alternative, namely corporate bonds.

No natural buyer existed for the enormous embedded optionality in the MBS universe, however. For that reason, Wall Street sought unnatural buyers. In part it solved the problem by selling the companion tranches, whose average life is extremely volatile, in the form of floating-rate instruments, on the theory that the value of such an instrument would deviate widely from par only in extreme scenarios. Japanese banks became major investors in such instruments during the early 1990s. Insurance companies bought companion "inverse floaters" (whose performance resembles that of a levered position in a fixed-rate companion bond) in order to add to yield, and to enhance duration in a falling interest-rate environment. In addition, the most option-laden tranches of the MBS universe (popularly qualified as "toxic waste") became the specialized province of MBS hedge funds, of which the largest was Granite Capital.

In other words, regulation led to a market inefficiency (namely, the excess offering of interest rate call options); inefficiency led to an innovative solution (the creation of the CMO); and the solution led to a new set of problems. These erupted in the spring of 1994 when the Federal Reserve raised interest rates after a prolonged period of low short-term rates. Granite Capital failed, along with some smaller MBS hedge funds, and a number of institutional investors incurred painful losses, notably Piper Jaffrey. Several insurance companies faced uncomfortably large losses in their holdings of CMOs, especially inverse floaters. CMO production shut down for the better part of a year. Complex CMO tranches traded at a deep discount to fair value. A salesperson would call a customer in late 1994 explaining that a certain inverse floater could be had at 40 percent below fair value, and the customer would respond, "Yes, I know that bond. My boss got fired for buying it."

An entire generation of MBS investors left the business. Market failures occur when existing market participants cannot buy the existing float, and are resolved when new market participants enter the market in order to take advantage of distressed prices. A new generation of MBS investors appeared in the form of hedge funds managed by former Wall Street MBS traders, attracting investment on the strength of their expertise in hedging the volatile securities. Again, the solution created the conditions for a new crisis. When the Federal Reserve again tightened interest rates during the spring of 1998, hedge funds heavily invested in MBS suffered distress. Disorder in the MBS market was aggravated by the Russian default in August 1998, leading to the celebrated failure of Long-Term Capital Management, which had invested heavily in both MBS and emerging market securities.

Once again, the composition of ownership of MBS changed. This time the government-sponsored enterprises (GSEs), namely the Federal National Mortgage Corporation ("Freddie Mac") and the Federal Home Loan Mortgage Corporation ("Fannie Mae") increased their holdings, along with large commercial bank treasury departments. At this writing just over half of the total universe of MBS securities is held by the GSEs and the commercial banks. Unlike traditional bond investors, the GSEs and the banks hedge the duration exposure of MBS using a variety of tools, mainly interest-rate swaps and options on swaps ("swaptions"). Federal regulators as well as private commentators have expressed concern about the potential exposure of the GSEs to violent swings in MBS valuation. I believe that the likelihood is quite remote that MBS volatility would exceed the management capacity of the GSEs, but we shall have to wait for another interest rate cycle with the shock value of a 1994 to know that for sure.

Regulatory Restrictions on Credit Quality and Index-Based Investing

During 2002, more than 3 percent of the nominal value of the U.S. investment-grade debt universe defaulted—the worst number since the Great Depression (see ***Figure 12.1***). Enron, WorldCom, and a handful of merchant power issuers accounted for most of that quotient. Associated with the high default rate for investment-grade debt was the first true market failure in the investment-grade credit market in two generations. Like all market failures I have observed, this one stemmed directly from a market inefficiency brought about by regulation, namely the requirement that pension funds and life insurance companies hold 95 percent of their debt portfolio in names of investment grade. From the vantage point of financial theory this restriction is senseless. If it were not, why should the same pension funds not have a requirement to own large-capitalization rather than small-capitalization equities, or for that matter, equities of firms that also enjoy an investment-grade debt rating? There is no evidence that speculative-grade debt offers worse performance over the long time horizon of pension-fund inves-

FIGURE 12.1 **Dollar-Weighted Investment-Grade Corporate Bond Default Rate**

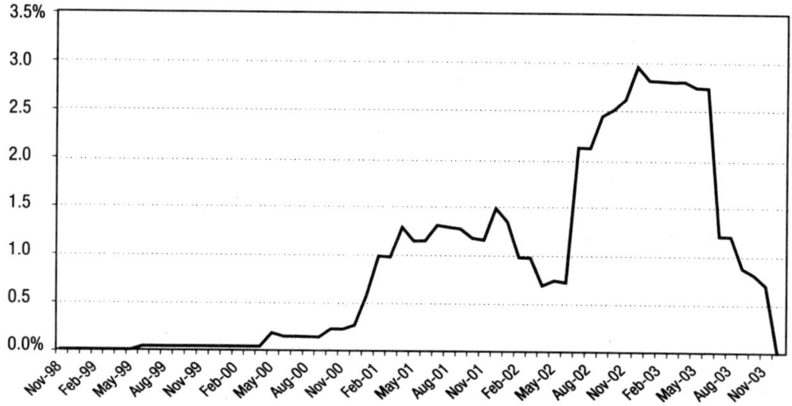

Source: Moody's Investor Services

tors than does investment-grade debt. Nonetheless, that is the rule, and it is unlikely to change in the foreseeable future.

Although Enron and WorldCom were special cases, they expressed in a somewhat more extreme fashion a governance problem that became rampant during the late 1990s and the early part of this decade. During the boom years of the equity market, corporate managers paid themselves in options. In consequence they had a long position in volatility, which is to say that they had much more to gain from the potential success of risky business strategies than they had to lose in the event of failure. In consequence, corporate America levered up its balance sheet in order to double down on its debts, while it placed bets on whatever fad seemed likely to attract the attention of shareholders—broadband, Internet content, and so forth. In fact, the majority of all corporate bond issuance between 1996 and 2002 was used to buy back equity, that is, to increase balance sheet leverage (see ***Figure 12.2***).

Why did investors agree to finance the buildup in leverage? Neither lack of intellect nor lack of information explains investor behavior in this regard; I am quite sure that this is the case due to personal observation. Instead, the practice of measuring fixed-income portfolio returns against an aggregate bond market index introduced a conflict between investors' long-term economic risk and their short-term business risk. As long as the ultimate beneficiaries of investment managers—that is, corporate pension funds—measure performance on the basis of a Sharpe

FIGURE 12.2 **More Than Half of All Net Corporate Issuance Since 1995 Funded Share Repurchases**

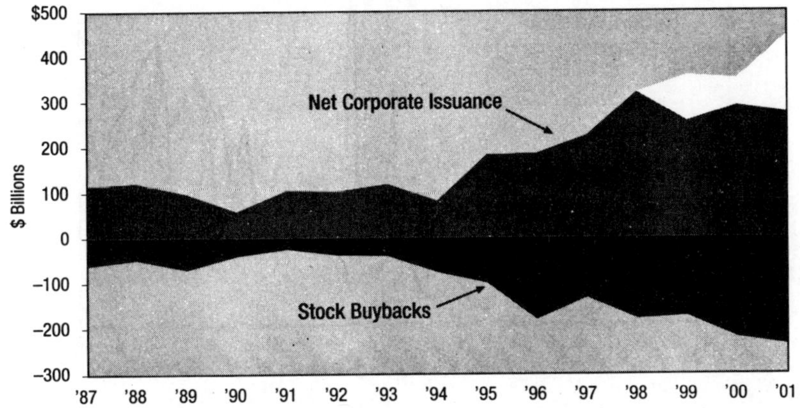

Source: Trimtabs

ratio calculated monthly, a decision to deviate from index weightings presents risk to the investment-management business. Corporate pension fund sponsors are highly risk averse; they tend to reward managers with additional funds for a slight degree of outperformance, and punish them drastically for underperformance.

The behavior of plan sponsors jars with the nature of corporate bond excess returns. If a manager exhibits superior credit judgment, her excess returns necessarily will be bunched into the small number of months in which credit events actually occur. Telecom bonds, for example, were the single best-performing sector in the investment-grade market during 2001, and WorldCom, whose fraudulent accounting had not yet been detected, was among the best performers. Consequently, it was a most popular name at the outset of 2002 among investment-grade managers. A manager who knew in late 2000 that WorldCom would fail, but did not know quite when, would have underperformed miserably during 2001, and lost funds under management to other managers.

Because the ratings process is backward-looking and tends to lag events, investment-grade managers accumulate names during a deteriorating credit cycle that later become subject to downgrade. The business risk of the money manager under the present regime of performance management causes managers to stick fairly close to index weightings. When the telephone sector flooded the market with new issues dur-

FIGURE 12.3 **Higher Leverage Led to Increased Volatility**

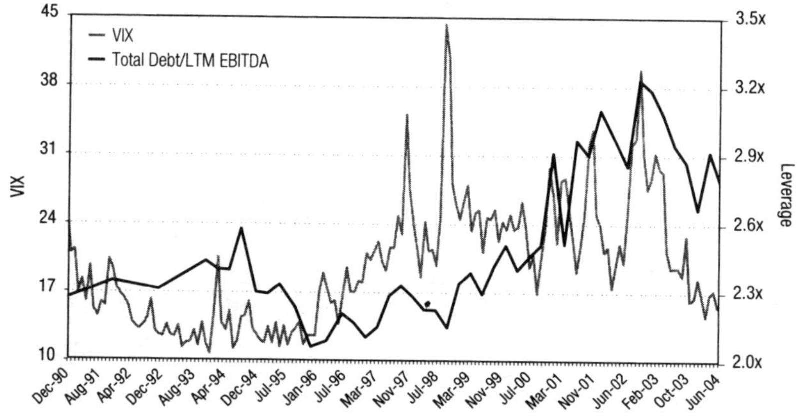

ing 2001 and 2002, most managers simply bought according to index weightings, despite widespread skepticism about the sector. By the time that the credit cycle reaches the point at which substantial portions of the universe are downgraded to speculative grade, it is too late.

Under the circumstances it is no surprise that the volatility of prospective business outcomes reached levels hitherto unknown (see *Figure 12.3*). The implied volatility of options on the S&P 500 equity index (the VIX index) reflected the underlying volatility of prospective earnings. VIX had traded in the low teens during the halcyon days of the mid-1990s, but averaged somewhere in the mid-20s during 1999–2002, spiking above the 40 percent range on several occasions (see *Figure 12.4*). A volatile business, by the same token, is more likely to suffer a ratings downgrade. In the wake of the WorldCom failure and the downgrade to speculative grade of most of the merchant power sector, corporate bond investors froze. By the summer of 2002, viable investment-grade companies had lost access to the capital markets. Because investment-grade managers must sell a name in their portfolio in the event that it descends to speculative grade, they will not buy a name in danger of a downgrade.

During the first week of August 2002, for example, Sprint bonds traded at 43 cents on the dollar, despite the fact that Sprint maintained a respectable investment-grade rating and was in no financial difficulty. Instead, investors feared a repetition of the mass reduction to

FIGURE 12.4 **VIX Index of Equity-Implied Volatility versus BBB Credit Spreads**

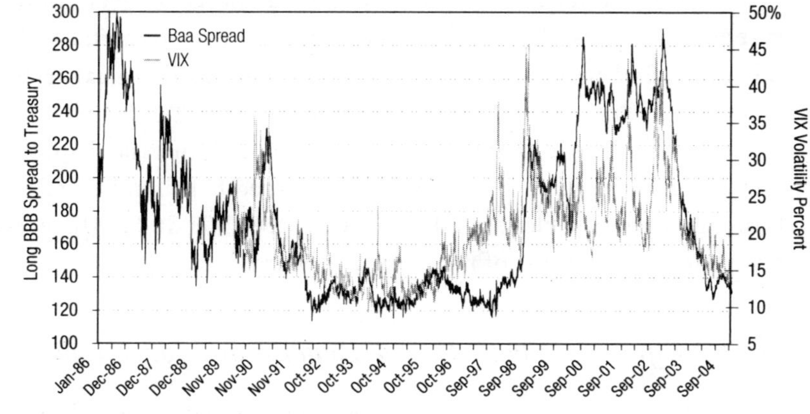

speculative grade that had occurred earlier in the merchant power sector and backed away from the market.

Once again, a new class of investors resolved the market failure, in this case the commercial banks. It had become an article of faith in the financial industry that the banks had been disintermediated out of the investment-grade market. Why would investment-grade borrowers tolerate loans with inconvenient covenants, as well as bank supervision, when the public market would lend them money with virtually no conditions? The patience of the public market gave out, of course, and the banks came back in. During the second week of August 2002, a bank consortium provided for Sprint's liquidity needs, and within a day or two of the announcement of a syndicated loan for the beleaguered issuer, its bonds once again were trading in the mid-70s.

Between October 2002 and the end of April 2003, the average spread to LIBOR (London interbank offered rate) of five-year U.S. investment-grade debt shrank from 180 basis points to only 110 basis points—the fastest rate of spread compression of which I am aware. By the same token, the average implied volatility of options on the equities of investment-grade issuers fell from a frightening 60 percent in mid-October 2002 to the low 30 percent range at the end of April 2003. Once the banks resolved the market failure by reentering the investment-grade market, public market investors considered the governance crisis at an end.

There appears to be little doubt that the corporate credit crisis of 2002–2003 contributed to the onset of recession as well as to the languid pace of recovery. Corporations invested too much during 1998–2001, saturating the economy with broadband and other technology. They required a pause in which to work off the excess capital stock, in real terms, and to rebuild their balance sheets, in financial terms. Credit cycles of this sort are long. Not since the early 1990s, when the real-estate market nearly brought some major commercial banks to their knees, had the U.S. economy suffered credit difficulties of this kind. In fact, there is a close parallel between the 1991 and 2001 recessions as well as the pattern of recovery afterwards.

The Shift to a Swap-Based Benchmark for Credit

During the spring of 2000, the spread between the ten-year U.S. Treasury note and the ten-year interest rate swap rate spiked to more than 140 basis points, from fewer than 70 basis points at the end of 1999 (see ***Figure 12.5***). Corporate investors who typically measure their returns against a benchmark index composed of Treasuries as well as corporates found that corporate spreads to Treasuries widened sharply along with swap spreads. The emergence of swaps rather than Treasuries as a benchmark for corporate debt threw a monkey wrench into the works of traditional corporate bond management.

FIGURE 12.5 **Yield Curve Flattening of 2000 versus Swap Spreads**
Weekly Observations 1/1/98 to 11/5/04

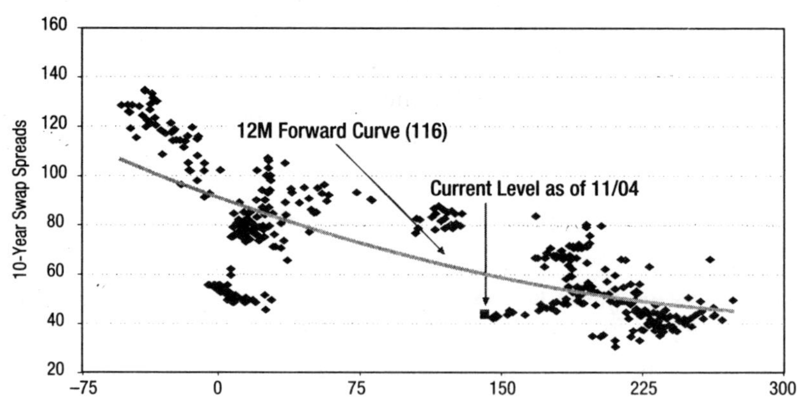

Source: Bloomberg

According to the textbook, the market for intermediate- and long-term corporate debt should mediate savings and investment over the same time horizon. Individuals who wish to retire in thirty years lend money to corporates who wish to acquire assets that depreciate over thirty years. In the remote past of, say, 1996, this model applied quite well. Life insurance companies and pension funds bought most new corporate debt issues. During 1999, however, the balance of power on capital markets swung toward levered credit buyers: commercial banks, bank-sponsored securities arbitrage conduits, semi-official agencies (such as German and Austrian *Landesbanken*), central banks, and so forth. Hedge funds emerged as a major factor in corporate bond markets during 2002.

This is quite different from the old maturity mismatch game, of course. Once upon a time, American thrift institutions borrowed from the public in the form of passbook savings accounts and lent the money in the form of thirty-year mortgages. The elimination of federal restrictions on short-term deposit interest at the outset of the 1980s coincided with a spike in short-term interest rates. The thrifts found themselves paying higher interest on their liabilities than they could earn on their assets, and the entire sector became insolvent. Today's financial institutions use the swap market to match floating-rate liabilities to floating-rate assets. In other words, they wish to pay fixed (sell swaps) to offset the fixed-rate payments they receive from

FIGURE 12.6 **Foreign Corporate Bond Purchases Made Higher Issuance Possible**

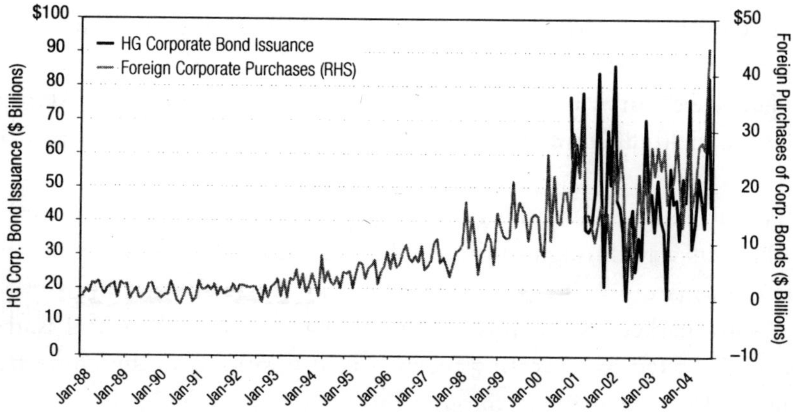

Source: Trimtabs; Federal Reserve Board

corporate issuers, and receive floating in order to pay their own floating-rate liabilities. As more and more leverage is applied to spread product, it is the swap market that bears all the pressure, and it is in the swap market that little crises of adjustment break out with unpleasant regularity.

Crises break out when supply and demand fall into imbalance and too many market participants wish to pay fixed. In August 1998 and August 1999, payers of fixed lined up at the swaps window out of fear (of systemic failure in the first case and of Y2K in the second case). At the end of January 2000, they wished to pay fixed because the relative income advantage for receiving fixed as opposed to floating had disappeared with the flattening of the yield curve.

With the best of intentions, the three principal official influences on capital markets—monetary policy, fiscal policy, and bank regulation—have conspired together to bring about the worst of all possible outcomes:

1. Monetary policy was attempting to exorcise the ghost of inflation, leading market participants to weight their bets in favor of falling future interest rates, thereby inverting the yield curve.

2. Fiscal policy was removing Treasuries from circulation, and the U.S. Treasury has led the market to believe that it will remove the longest-term debt from circulation fastest.

3. Regulatory policy required dealers to calculate their capital according to a Value-at-Risk model that compels them to reduce exposure in periods of market volatility. In short, dealers could not provide sufficient liquidity to make swaps an efficient benchmark.

The following sequence of events caused the spring 2000 distress in the swaps market:

1. The yield curve inverted in anticipation of Fed tightening, starting from the yield differential between maturities of from ten to thirty years and moving toward the front.

2. As the long end of the coupon curve rallied, the inversion began working its way through to the yield differential between two- and ten-year Treasuries.

3. Swaps market participants began selling ten-year swaps—that is, trying to pay fixed (or reversing receive-fixed positions) as the curve flattened—eliminating the economic incentive for receiving fixed in the first place.

4. Faced with a long line of customers seeking to pay fixed, dealers scrambled to obtain on-the-run Treasuries with which to hedge swap positions in the longer swap maturities.
5. On-the-run Treasury prices gapped relative to off-the-run prices, while the repo rate for on-the-run ten-year notes fell to a zero handle.
6. Dealers aggressively widened their bid for ten-year swaps, pushing the ten-year swap spread out to a late-April level of +146 basis points (at which level, to be sure, little execution took place).

The derivative market outguns the Treasury market. It is becoming hard to tell whether disarray in the derivatives markets whipsawed the Treasury market or vice versa. The Treasury's confusion about potential buyback programs was a lit match, to be sure, but the match had the misfortune to land on a gasoline spill.

Multiple choice: In the future, corporate spreads will be benchmarked against
1. on-the-run Treasuries
2. off-the-run Treasuries
3. swaps
4. agencies
5. other corporate debt
6. all of the above
7. none of the above

As far as I can tell, both (6) and (7) are full credit answers. There really is no benchmark for credit markets, and enduring confusion over how to price corporate debt will make the credit markets susceptible to the kind of turbulence the markets endured in the spring of 2000. Swaps rather than Treasuries have become the "benchmark" for corporate debt, in the same way that Lucky Strikes became a currency in Germany after April 1945. Lucky Strikes, of course, never were intended to be a currency, did a poor job of acting like a currency, and ceased to be a currency as soon as circumstances changed. Nonetheless, American cigarettes were for a time the closest things to a currency that Germany possessed. Lucky Strikes became the German currency because American troops occupied Germany and could provide liquidity in the form of cigarettes. Swaps have become the benchmark for credit markets because LIBOR-based credit buyers set prices at the margin for liquid, higher-rated corporate product.

U.S. agencies buy mortgage-backed securities, that is, sell convexity in the form of embedded call options. They repurchase the convexity they sell in the MBS market by purchasing swaptions (options to enter into swap agreements). Whether the agencies swap proceeds of their bullet bond issues into floating, or buy swaptions giving them the option to do so at a later date, agency supply has a direct influence upon swap spreads.

What makes swaps problematic as a benchmark is the market's attempt to use them as a universal hedging instrument. As spreads themselves become more volatile, dealers, investors, and issuers increasingly use swaps as the preferred hedging vehicle. This phenomenon was also reinforced after the LTCM crisis in the fall of 1998, as many levered speculators in the swap market were driven out. When the swap market itself becomes volatile, it becomes difficult to hedge with swaps. In this situation, we expect swap dealers to quote higher swap spreads to compensate for the additional risk they are taking. In other words, in a swap market dominated by hedging activities, swap spread and swap volatility should be positively correlated.

13

Europe's Political and Economic Future

THOMAS MAYER

THOMAS MAYER has seen the economic experiment that is Europe, and he does not like the view. That he is Deutsche Bank's chief European economist is enough to give one pause. Mayer writes on economics with such authority, such command of its linkage to politics and history, that he deserves our immediate attention. Mayer addresses the limited set of combined fiscal and monetary options now made available to an aging Europe. No rose-colored glasses for this former International Monetary Fund economist. Here is Thomas Mayer with a clear-eyed and sober vision.

The "historicist" school of philosophy (remember Hegel and Marx?) claimed that they had found the key drivers of human history. They liberally used their historical model to predict the future course of events (culminating in the triumph of the Prussian state or the advent of communism, as the case may be). But even before history proved them wrong, Sir Karl Popper proved that "historicism" was a fallacy.[1] Today, we have come to accept that it is beyond our capabilities to predict the course of history. All we can do is identify a few forces shaping the recent past and possibly prospective developments, and speculate where they may lead us. It is as if we were looking down a street: We may see ahead to the next corner—no more, but also no less.

What do we see when we look down the street of European political and economic history? For centuries, Europe was the economic and political powerhouse of the world. It owed its rise to a favorable climate, a culture of recognizing private property, and the good luck of being spared a strong central political power after the fall of the Roman Empire.[2] As a result, there was intense economic and political competition among numerous peoples living in a rather small geographic area. What we often regard as the "dark middle ages" between the fall of Rome and the beginning of the Renaissance period in fact marked the incubation time for the superpower that dominated the world between the sixteenth and the beginning of the twentieth century. In the words of historian Paul Kennedy, "the political and social consequences of this decentralised, largely unsupervised growth of commerce and merchants and ports and markets were of the greatest significance."[3]

At the end of the nineteenth century Europe had reached the peak of its power. It accounted for almost 50 percent of world manufacturing output, and its production of iron and steel exceeded that of the United States, Russia, and Japan combined. But what may have been productive rivalry and competition among European peoples in previous times turned in the twentieth century into a disastrous struggle among European nations. By 1938, halfway through this struggle, Europe's share in world manufacturing output had been overtaken by the United States and the rest of the world (see ***Figure 13.1***). In 1945, when the hot phase of the conflict ended, Europe was divided between a United States and a Russian sphere of influence, and by 1989, when the end of the Cold War finally concluded Europe's twentieth century war, the United States was left as the only remaining world superpower.

The fall of the Berlin Wall brought hopes of a European revival. Europe had shaken off the deep divisions that had caused two world wars and the Cold War; it now seemed free to catch up with the United

FIGURE 13.1 **Shares in World Manufacturing Output (%)**

	1880	**1900**	**1913**	**1928**	**1938**
Europe*	46.1	45.7	41.3	30.2	30.6
United States	14.7	23.6	32.0	39.3	31.4
Rest of world	39.2	30.7	26.7	30.5	38.0

* Territory of the first fifteen member states of the European Union.

Source: Paul Kennedy, *The Rise and Fall of Great Powers* (New York: Vintage, 1987), p. 202.

States and become its equal partner in a prosperous world. But Europe continued to fall behind both politically and economically during the 1990s. It is time now to recognize that this was not just an aborted take-off to be repeated at a later time. Rather, developments during the 1990s confirm the relative political and economic decline that Europe has been experiencing since the beginning of the twentieth century. Since Europe is unlikely to meet the coming political and economic challenges, decline is set to continue for at least another two to three generations.

The Political Challenge

Henry Kissinger once quipped that Europe had no one telephone number other political leaders could call when needed. Clearly, to be taken seriously in global politics, a country or region needs to show the appropriate weight on the economic and/or military scale, and it needs to be able to use this weight effectively. In recognition of the laws of "power politics," European politicians have created political institutions that may be "called" by other world leaders. For a while, the process of European integration seemed destined to end in a fully fledged political union, encompassing a population larger than that of the United States and a GDP second only to U.S. GDP. However, with the fall of the Berlin Wall and the end of Europe's post–World War II political division, the drive toward European political integration has stalled. It now seems clear that for the foreseeable future Europe is unlikely to acquire a single telephone number, and the world will feel that it can manage without it.

In Search of Political Stability

The idea of "ever closer union," anchored in the 1957 Treaty of Rome, the cornerstone of the European Union (EU), owed its birth to the lessons drawn from World War I. Already before the end of World War II, the Allies made up their mind not to repeat the mistakes made after their victory in 1918. Then, a politically unstable Germany was saddled with large reparation claims and left to itself.[4] Poor economic performance and resentment against war reparations created the breeding ground for the Nazi Party and Germany's renewed aggression against its neighbors. This time, a defeated Germany was not to be left alone. It was to be assisted to reconstruct its war-torn economy and to be integrated into a European political structure. As early as 1944, Robert Marjolin, one of the architects of European unification and in exile in Washington, D.C., at the time, developed plans for European political integration. "The first stage would be to form a federation comprising Britain, France, Benelux and Germany," he recalled in his memoirs.[5] He also expected an extraordinarily powerful impetus from the unification of the European market, which had to include the dismantlement of all barriers to the free movements of goods, people, and capital.

Marjolin's views were strongly supported by the U.S. administration and Congress. The Americans saw integration as essential for Europe to be able to stand again on her own feet economically and politically without requiring open-ended U.S. economic and military support. However, despite Winston Churchill's musings about European political unification in 1940, the United Kingdom was hostile to the idea. While war had severely disrupted the Continent—destroying economies and discrediting governments—Britain came out of it without "a sense of national failure and a feeling of national inadequacy but with a sense of national achievement and cohesion and an illusion of power," according to historian Miriam Camps.[6] Hence, there was no emotional support for European unity. Rather, Britons still felt that their country was the third power after the United States and Soviet Union, and that they enjoyed a special relationship with the United States.

As a result, European integration started without the United Kingdom. For the key French policymakers of the time—Charles de Gaulle and Jean Monnet—the improvement of relationships between France and Germany was the cornerstone of European unification. Monnet started the process by concluding the treaty establishing the European Coal and Steel Community in 1952,[7] and de Gaulle later carried it further in a plan for an eventual European Economic Union, in the Treaty of Rome. The French policymakers' view of the need to reconcile the two countries was fully shared by their German counterparts. For the latter, France provided the political and diplomatic platform allowing the young Federal Republic of Germany to begin its return to international respectability. The price was accepting France's leading role in Europe, and it seemed a small price to pay for German chancellors from Konrad Adenauer to Helmut Kohl.

For the Germans—and other "federalist" Europeans—European integration followed the principle of chain reaction.[8] Starting out with a customs union and a common agricultural policy, it would inevitably lead to a common monetary, foreign, and defense policy. At the end

Timeline

1951	Treaty of Paris establishes the European Coal and Steel Community.
1957	Treaty of Rome establishes the European Community (EC). Member states include Belgium, Germany, France, Italy, Luxembourg, and the Netherlands.
1973	Denmark, Ireland, and the UK join the EC.
1981	Greece joins the EC.
1986	The Single European Act sets a timetable for establishing a "common market"—that is, allowing the free movement of goods, people, services, and capital throughout the EC by 1992. Spain and Portugal join the EC.
1992	The common market is achieved on schedule. The Treaty of Maastricht sets a timetable for Economic and Monetary Union (EMU). The European Community adopts the designation European Union (EU).
1995	Austria, Finland, and Sweden join the EU.
1999	Eleven member states meet key criteria for EMU. These eleven countries—Austria, Belgium, Germany, Finland, France, Ireland, Italy, Luxembourg, the Netherlands, Portugal, and Spain—are referred to as "Euroland." The euro is launched. (Note that Sweden has not met the exchange-rate criterion for EMU membership and that Denmark and the United Kingdom have opted not to adopt the euro. Norway and Switzerland opt not to belong to the EU at all).
2001	Greece joins EMU, bringing the count of Euroland states to twelve.
2004	Ten countries join the EU—Cyprus, Czech Republic, Estonia, Hungary, Latvia, Lithuania, Malta, Poland, Slovakia, and Slovenia. EU heads of states and governments sign the European Constitution.

stood a politically united federal Europe. The pragmatists—including most of the French political elite—did not believe in the United States of Europe, but nevertheless saw a process underway where European nation states would cooperate in an increasing number of areas on the basis of international treaties rather than political union. For many years, this difference in view did not matter. Both the federalists and pragmatists could push for the same policies of "ever closer union" without having to sort out where it would end.

European Monetary Union as a Stepping Stone to Political Union

For a long time, monetary union has been seen as a crucial step toward "ever closer union" and a possible catalyst for political union.[9] When customs union and the Common Agricultural Policy were established at the end of the 1960s, the European Council, in December 1969, called for establishment of a European Monetary Union (EMU). An expert group, led by Luxembourg prime minister Pierre Werner, was set up with the mandate to produce a plan for achieving EMU. In October 1970, the group delivered its blueprint, the so-called Werner Plan, which foresaw establishment of EMU within a decade—that is, by 1980. However, the break-up of the Bretton-Woods System in 1973 and the ensuing turmoil in the foreign exchange markets derailed the plan. Nevertheless, one element of the plan survived: a mechanism of quasi-fixed exchange rates among key European currencies, the so-called "snake." This mechanism—which existed from 1972–1978—provided the basis for the European Monetary System that was launched in 1979.

In the following years, the objective of creating EMU faded into the background. Economic policies within Europe and at the G7 level drifted apart, and exchange rates exhibited large fluctuations. France and Italy continued to favor progress toward monetary integration—in part because they resented the dominant position of the deutsche mark in Europe—while Germany resisted it. The German currency had become a symbol for Germany's economic miracle in the wake of World War II, and the German population was extremely reluctant to bring it into a common currency with countries that had dubious records of price and currency stability.

However, in a surprising response to French and Italian criticism of the "asymmetric" nature of the EMS (which they felt had in fact established a deutsche mark standard in Europe), German foreign minister

Hans-Dietrich Genscher in 1988 proposed to establish a European Central Bank (ECB). He argued that a common currency and central bank would act as catalyst to achieve economic convergence among the member states, which was needed to establish a sound basis for any stable monetary arrangement in Europe. In a follow-up memorandum, German finance minister Gerhard Stoltenberg spelled out the prerequisites of EMU from the German perspective: an independent central bank, committed to price stability, and free movement of capital. Given that the German public—and the powerful Bundesbank—were deeply skeptical about EMU, it is not entirely clear what prompted German government officials to take the initiative. Perhaps they thought that their proposals were too far-reaching for the French and Italians, and a rejection of the German initiative by these countries would end the nagging complaints about the "asymmetry" of the EMS. However, despite considerable scepticism, the European Council at its meeting in Hanover in 1988 established a committee led by European Union president Jacques Delors to study and propose concrete steps toward monetary union. The committee's report ("Delors Report"), presented in early 1989, became the nucleus of the 1991 Maastricht Treaty, which set a timetable for achieving EMU by the end of the decade.

Perhaps the new initiative for the creation of EMU launched in 1988 would have suffered the same fate as the earlier Werner Plan had the Berlin Wall not fallen in 1989. Chancellor Helmut Kohl realized sooner than many others that this event paved the way for German unification, and to assuage the fears of his nation's neighbors about a reunited Germany, he was willing to anchor Germany even more firmly in the EU. For France, the best assurance of irreversible integration of Germany into Europe was monetary union, and Kohl accepted the price of surrendering the deutsche mark to receive French support for unification. German unification and the creation of EMU were closely related events. Perhaps the former would have happened even without the latter, but it is rather doubtful whether the German public would have given up its beloved deutsche mark for anything less than unification.

Momentum Toward Political Union Evaporates

With the fall of the Berlin wall, German unification, European monetary union, and the return of Central European countries to Europe (resulting in their membership in the European Union), Europe's twentieth-century war was finally concluded. But with it, the post–World War II drive for political European integration lost its momentum. In a last effort to drive political union forward, Chancellor Kohl and President Francois Mitterrand, at the Dublin European Council meeting of April 1990, decided to convene an Intergovernmental Conference (IGC) on political union to proceed in parallel to the IGC on monetary union. However, the results of this initiative were disappointing for anyone favoring full political integration.

The conclusion of the IGC on political union at the Amsterdam summit of 1997 brought only limited further integration in the area of social and employment policy and, most important, adoption of the Stability and Growth Pact (SGP) as a means to coordinate fiscal policy in EMU. The German side, in particular, regarded the SGP as a (second best) alternative to a common fiscal policy in a political union. After the conclusion of the Maastricht Treaty, senior German representatives, including Bundesbank president Hans Tietmeyer, had argued that EMU would not be possible without political union.[10] Having failed to achieve the latter, they looked for other measures to fortify EMU. If political coordination of fiscal policy in a federal system was not feasible, perhaps a simple and mechanic rule could impose some fiscal discipline. The last effort at greater political integration—the draft of a European Constitution completed in 2003—brought little more than a consolidation of existing EU treaties and streamlining of member states' voting rights in an enlarged EU. Although governments have agreed on it in 2004, ratification of the constitution in twenty-five national parliaments looks highly uncertain.

Political cooperation in certain areas notwithstanding, Europe is likely to remain politically fragmented in numerous small to medium-sized nation-states. Such a Europe is always vulnerable to outside influences and prone to political division and centrifugal forces (as recently exhibited in the conflict about the Iraq war). It is hard to see how such a Europe cannot lose political weight on the world stage against the United States and an increasingly more important China.

The Economic Challenge

The European economic model is a market economy with a strong role for the government, which establishes rules for goods, services, labor, and capital markets; manages macroeconomic policy; and pursues a proactive social policy. In the German version of the model—the so-called social market economy developed by Ludwig Erhard (among others) during World War II and implemented in the post-war Federal Republic—the government ensured the functioning of the market through competition policy and corrected the income distribution resulting from market processes through social and tax policy. In the French version ("planification"), the government initially took an even more active role, establishing multiyear economic plans with concrete objectives for private and nationalized industries. In Italy, the government exerted its influence on the economy through state ownership in a large part of the industrial sector. In all European countries, the agricultural sector has been under intense government management through the Common Agricultural Policy of the EU.

Over time, direct government meddling in industry has receded. Following the British example of the early 1980s, continental European governments—especially Italy and France—have privatized a large part of their extensive industrial holdings. Moreover, industrial policy (e.g., in the context of French "planification") has become less intrusive. At the same time, however, governments have continued to redistribute income and fortify the welfare state. Hence, government spending in general and spending on social benefits in particular have remained much higher relative to GDP in Europe than elsewhere, notably in the United States (see ***Figure 13.2***).

The strong role of government and the comprehensive social security afforded individuals has over time introduced considerable distortions in the European economy. Among other things, vested interest groups intent on seeking and defending economic rents have become powerful players, distorting goods and factor prices. Individuals have

FIGURE 13.2 **Social Spending (as % of GDP, 1995–1998)**

France	**29.1**
Germany	**27.4**
Italy	**24.5**
United Kingdom	**25.4**
United States	**15.1**
Canada	**18.4**

Source: OECD

developed considerable moral hazard, relying on the welfare state to care for them during periods of sickness, unemployment, and old age, instead of building up private savings and taking other precautions on their own. Moreover, attitudes toward risk-taking in economic life have changed, with individuals preferring low risk and less demanding activities to entrepreneurship and self-employment. These behavioral changes have gradually undermined the longer-term viability of the European economic model. In the early 1980s, the United Kingdom broke with the European economic model and adopted a more market-oriented approach. But continental European countries have chosen to adhere to it. The model now faces serious challenges from monetary union, globalization, and demographic change.

Europe's Economic Model and EMU

The Maastricht Treaty set "nominal" convergence of inflation rates, interest rates, government deficits, and debt ratios as a prerequisite for the entry of a country into EMU. "Real" convergence of activity and employment growth was expected to follow, so that a common monetary policy would be suitable not only for the Euroland aggregate but also for individual member countries. Although fiscal policy was still conducted at the national level—and hence could be tailored much better to individual countries' needs—its room for maneuver was significantly limited in most countries by the need to bring government deficits down to levels sustainable in the medium to long run. The Stability and Growth Pact was seen to ensure that countries indeed adhered to fiscal policy prudence.

With macroeconomic policy lifted to the supranational level or subjected to rules, and microeconomic flexibility limited, a smooth functioning of EMU would have required a synchronization of cyclical developments among euro area countries (i.e., "real convergence"). However, developments in economic activity continued to differ among EMU member countries, and the stance of macroeconomic policy determined by the ECB and the fiscal policy rules did not fit all countries equally. Something had to give.

A few economists had argued that tensions within EMU would eventually crack structural rigidities. In other words, macroeconomic inflexibility was seen to lead to microeconomic flexibility—a partial break from, or at least a serious transformation of, the European economic model. Others had argued that a "one-size-fits-all" monetary policy in a diverging union would create resentment against the ECB and could eventually induce some countries to pull out of EMU. So far, these predictions have not come true. Instead, the economic pressure resulting from diverging real economic developments in a microeconomically relatively rigid union has affected fiscal policy. Countries have severely fallen behind the objective of balancing their structural budget balances, and some countries have been unable to keep their deficits below the 3-percent-of-GDP ceiling (established in the Maastricht Treaty as the upper limit for acceptable deficits in EMU). Political pressure has been applied to shield fiscal policy offenders from the consequences of their violations. Thus in November 2003, ECOFIN (the Council of the EU meeting at the level of economic and finance ministers) suspended the excessive deficit procedure against Germany and France, which had faced imposition of a fiscal-adjustment program by the commission and eventually sanctions for noncompliance (see ***Figure 13.3***).

The break of fiscal policy rules raises two problems. First, it undermines the institutional basis of EMU. If provisions of the Maastricht Treaty concerning fiscal policy can be violated, how safe are then other provisions—for instance, the one guaranteeing the ECB independence from political interference? Second, continuing structural deficits and rising debt-to-GDP ratios now will exacerbate the fiscal stress in the

FIGURE 13.3 **Euro Area Stability Programs***

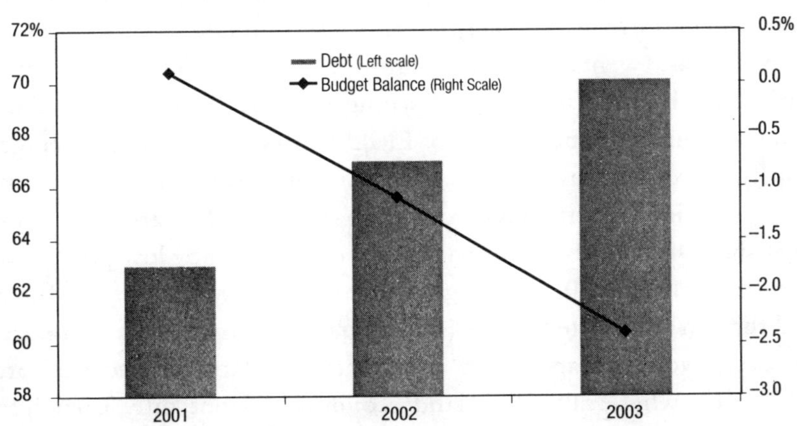

* Projected 2004 budget balance and debt (in % of GDP).

Source: European Commission, Economic and Financial Directorate

future when a smaller working age population will have to sustain a larger number of pensioners. The tax burden on workers to finance public pension benefits would then be raised further by the need to service a larger public debt (a significant part of which may be owed to pensioners and foreign residents).

The strains on EMU created by the microeconomic inflexibility of the European economic model are unlikely to lead to a political or economic break-up of the European Union (or its different institutions, including EMU). The political costs of such a break-up would be far higher than any conceivable economic gain from the unwinding of malfunctioning EU institutions. But they may well pour sand into the European economic machine. Persistent fiscal deficits will exert upward pressure on interest rates and private household savings rates (as people lose confidence in public old age benefits). This will depress investment, consumption, and, in the event, GDP and employment growth.

Europe's Economic Model and Globalization

The integration of lesser developed and formerly communist countries into a world market of goods and services, the fall of the barriers to international capital flows, and rapid technological progress led largely by the United States have put additional strain on the European economic model. Although some reforms have been started, the economic policy response so far has been inadequate. As a result of adjustment proceeding at a snail's pace and falling short of the requirements, European economic performance is likely to remain second rate.

In the past twenty years, GDP growth in the euro area has slowed markedly while it has remained strong or picked up in other major countries. Since the beginning of EMU in 1999, growth has averaged only 1.8 percent, compared with 2.8 percent in the United States and 2.5 percent in the United Kingdom (see ***Figure 13.4***). Part of the more sluggish GDP growth in Euroland can be explained by low population growth. In 1990–2002, GDP growth per capita was higher in Euroland than in the United States (see ***Figure 13.5***). In more recent years, however, even per capita GDP growth in Euroland fell short of rates reached elsewhere. The Euroland economy did not fully participate in the new economy boom toward the end of the 1990s, but it was similarly affected by the economic slowdown that followed the burst of the Internet stock bubble. Moreover, economic recovery from the

FIGURE 13.4 **Real GDP Growth, 1999–2003**

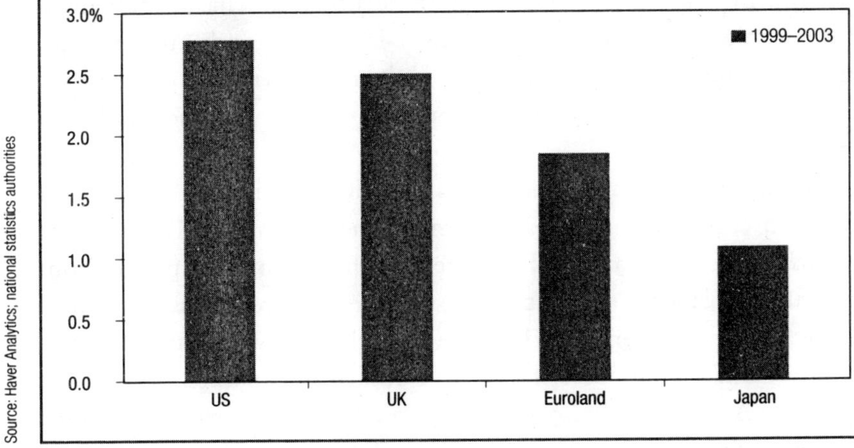

FIGURE 13.5 **Growth of Real GDP per Head, 1996–2003**
(5-year trailing moving averages)

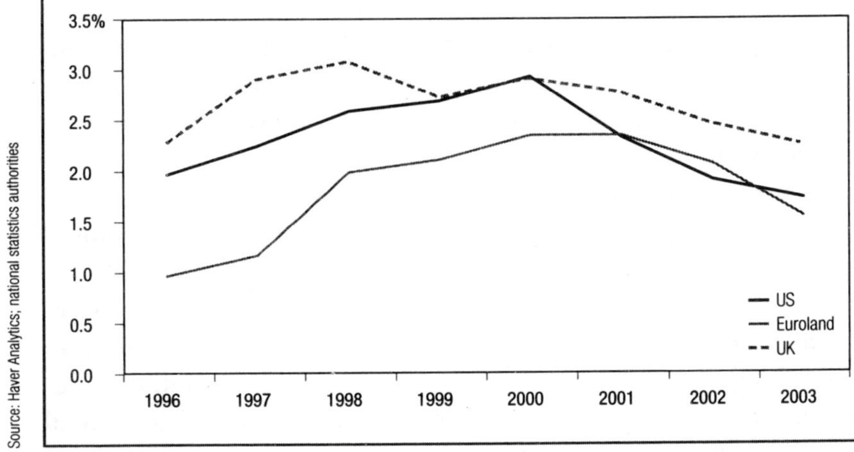

weakness of 2001–2002 was more sluggish than in most other industrial countries.

Difficulties to adjust to more intense international competition have been an important reason for Euroland's lackluster performance in recent years. The emergence of China and Southeast Asian countries as major suppliers of manufactured goods and the accession of Central European countries to the EU have increased low cost competition for European producers. With Euroland wages sticky, non-wage labor costs high, and labor mobility very limited, low-wage competition from abroad has put upward pressure on unemployment. Moreover, new information and communication technologies have challenged Europe's role as producer of technology-intensive goods and services. But government regulations, outdated work practices, and management deficiencies have slowed the application of the new technologies in the business sector.

Last but not least, increased international capital market integration has raised competition for capital. Since the end of the 1960s, firms have substituted capital for labor to reduce the pressure on profits from rapidly rising labor costs. Labor substitution was made easier because the costs of capital were kept low due to high domestic savings and the pressure from shareholders to raise the return on invested capital was moderate. A considerable part of industry was in government hands, and crossholdings of shares by companies was pervasive. But things changed with the privatization of government holdings in industry and the international integration of capital markets. More attractive returns to capital elsewhere caused capital outflows and deprived the business sector of its source of cheap funding. To raise their attractiveness to investors, Euroland companies had to restructure. In many cases, this involved paring down the capital stock to raise capital productivity. As a result, investment growth slowed from past levels and relative to rates achieved abroad.

Because of these developments, Euro area annual capital stock growth slowed from about 3 percent during 1981 to 1990 to only 2 percent in 2003. Partly offsetting this, total hours worked stagnated in recent years after having fallen significantly during the 1970s. However, total factor productivity growth—which reflects the economy's ability to leverage growth of capital and labor input through technology—fell by about one half since the mid-1990s. By contrast, total factor productivity growth in the United States almost doubled during this period.[11] Without a sizeable increase in productivity growth, it is hard to see how Euroland trend growth can pick up to its past levels and narrow its gap with the United States. In a recent study, Deutsche Bank estimated Euroland potential growth at 1.75 percent in the annual average of 2003–2010.[12] This estimate was based on a contribution from total factor productivity growth of 0.6 percentage points, from capital input growth of 0.7 percentage points, and from labor input growth of 0.5 percentage points. For this to materialize it was assumed that productivity and capital input growth stabilized slightly above recent levels, while labor input growth accelerated on the back of working-time extension, higher participation rates, and lower structural unemployment. Without an improvement in the labor market, potential growth was seen at only 1.5 percent.

Over time, the euro area's low trend growth will reduce its economic (and as a result also its political) weight in the world. In 2003, Euroland GDP was about 74 percent the size of U.S. GDP (when compared using purchasing power parities). If average annual euro area GDP grows by only 1 percent from 2003 to 2010, while U.S. GDP rises by 3.5 percent per year during this period, the size of the Euroland economy would decline to about 65 percent of the U.S. economy by 2010.

Europe's Economic Model and Demographic Change

The biggest challenge to Europe's economic model emanates from demographic change. With the population aging rapidly, potential GDP growth will decline while public pension and health systems will come under more financial strain. Fights over the income distribution between the working-age and retired generations may revive inflation. Higher interest rates and a decline in aggregate national savings are likely to erode the value of real and financial assets.

Fertility rates peaked in Europe in the mid-1960s, about ten to fifteen years after the United States. Since then, they have declined rapidly. In Germany, only 138 babies were born on average per 100 women in 2000, down from more than 250 babies around 1965. In the fifteen original EU member states, the fertility rate fell to 148, not much above the German level. With a little more than 200 children per 100 women needed to keep a population stable (without immigration), the drop in fertility rates heralds a decline in the European population in the future. The upcoming demographic change will have a number of implications.

First, potential GDP growth will decline. For a while, a lack of population growth can be compensated for by higher participation in the workforce and longer (weekly and lifetime) working hours. However, when the reserves are exhausted and the population begins to shrink, labor input will decline. For Germany, this is estimated to occur as of 2020, when the baby boomers will begin to retire. A higher capital-to-labor ratio can mitigate the effects of shrinking labor input growth on potential GDP growth, but an aging population may also suffer lower total factor productivity growth as innovation and adoption of new technologies fade.

Second, the aging of the population will raise upward pressure on government spending and depress the tax base. According to OECD calculations, net public debt in a "stylized" country could at least double relative to GDP due to the costs of aging during the first half of this century.[13]

Third, when demands for government services can no longer be reconciled with governments' ability to raise taxes and borrowing, inflation may increase. Historically, periods of fiscal distress were associated with high inflation as governments took recourse to the "inflation tax" as revenue of last resort.[14] They may need to do this again as government finances will probably be ill prepared to deal with the costs of aging. Higher inflation will not only quietly raise taxes through "bracket creep" but also erode the value of entitlements that cannot be cut in absolute terms for political reasons. Confronted with the choice of condoning economic weakening and the unravelling of government finances owing to intergenerational fights over the distribution of income or of accepting higher inflation, even an independent central bank may well choose the latter as the lesser evil.

Fourth, asset prices are likely to decline. With public pension benefits becoming less generous, people have to save more during their working lives to secure living standards during retirement. As the ratio of workers to pensioners declines in future years, the balance between saving and dis-saving will deteriorate. This, together with higher interest rates on the back of rising government borrowing and inflation, is likely to erode the value of real and financial assets. While a decline in asset prices is likely in all aging societies, it will be most pronounced in Europe—and especially in Italy and Germany—where the ratio of workers to pensioners will fall the most.

Riding into the Sunset?

In his impressive economic history of the world, *The Wealth and Poverty of Nations,* David Landes traces the rise and fall of several European powers since the middle of the last millennium. A number of factors generally came together to cause a nation to lose its economic edge: saturation and complacency in the wake of sizeable economic gains, failure to adapt an outdated "business model" to changing circumstances, and/or an intolerant government suppressing research, development, and the application of new technologies. None of the former European superpowers—including Spain, Portugal, Holland, or the United Kingdom—is likely to rise again to its earlier power and glory. Could it be that Europe as a whole suffers the same fate of irreversible relative decline in the world?

Perhaps. But economic history also holds the example of a country presently reversing its long-term decline: China. By the middle of the last millennium, China was militarily and economically more powerful and technologically more advanced than Europe. But a complacent government dominating all spheres of political and economic life pursued an inward-looking policy, causing China to lose its technological advantage and economic and military superiority. Only with the government's gradual withdrawal from interference into all spheres of life almost 500 years later has China started the process of catching up with the more prosperous parts of the world.

History may not repeat itself but, as Mark Twain observed, it sometimes rhymes. Internal wars, which have rocked the world, have undermined Europe's political standing in the twentieth century. Complacency, inward orientation, and heavy reliance on governments to deal with all vagaries of life have set its economy on a path of relative decline. However, like China, Europe need not be down and out forever. But whether the readers of these lines will live to see Europe's economic revival is an entirely different matter.

CHAPTER NOTES

1. Karl Popper, *The Poverty of Historicism* (London: Routledge, 1960).
2. David S. Landes, *The Wealth and Poverty of Nations: Why Some Are So Rich and Some So Poor* (New York: W.W. Norton, 1998).
3. Paul Kennedy, *The Rise and Fall of Great Powers* (New York: Random House, 1987), p. 19.
4. When Germany was unable to make the reparation payments, France in 1923 occupied the Ruhr region, Germany's industrial heartland. Different payment plans were worked out, with the Young plan of 1929 envisaging German war reparation payments until 1988.
5. Robert Marjolin, *Memoirs 1911–1986* (London: Weidenfeld & Nicolson, 1989), p. 127.
6. Miriam Camps, *Britain and the European Community* (Princeton University Press, 1964), p. 3.

7. It was no accident that European unification started with steel and coal. Being essential for armament, German coal and steel production had been subject to allied control. Bringing this industry into a community of states consisting of France, Germany, Italy, and the Benelux countries was equivalent to placing arms production in common hands.

8. Walter Hallstein—a close aide to Chancellor Konrad Adenauer and the first EEC president—regarded this as the "inner logic" of the European integration process. See Walter Hallstein, *Europe in the Making* (New York: W. W. Norton, 1972).

9. According to Jacques Rueff, an economic policy adviser to the French president Charles DeGaulle, "Europe will be borne through its money, or it will not be borne at all." For a brief history of European monetary integration, see Daniel Gros and Niels Thygesen, *European Monetary Integration* (New York: Longman, 1998).

10. Hans Tietmeyer, The relationship between monetary and political integration. In: A. Bakker, H. Boot, O. Sleijpen, and W. Vanthoor (eds.), Monetary Stability Through International Cooperation. Dordrecht 1994, pp. 21-30.

11. See Olivier Blanchard, "The Economic Future of Europe," NBER Working paper 10310, February 2004 and the studies cited there.

12. See "Euroland: Assessing Potential Growth," Deutsche Bank Global Markets Research Special Publications, October 1, 2003.

13. See OECD, *Economic Outlook,* No. 69, Paris, 2001, pp. 160-62.

14. See "Inflation Is Dead! Long Live Inflation!" Deutsche Bank Global Markets Research Special Report, April 2004.

14

The Economic Future of Asia

ROBERT D. HORMATS

ROBERT HORMATS writes our final, long chapter. Asia, India included, is the story of this twenty-first century. He reaches across Asia, across its countless time zones. Hormats, vice chairman of Goldman Sachs (International), writes with a holistic grasp that makes him an important source for political and economic leaders worldwide. Under crisis, he is one they call upon. Here, Hormats paints an optimistic canvas of an ascendant Asia. He believes in market-based solutions to the clash and turmoil of nation, culture, and economics, but also ones that address the region's social problems of still massive poverty and joblessness. But notice, within Hormats's vision, his regard for our need to expect the unexpected.

After several years of subpar growth, much of Asia is enjoying a period of stronger economic performance. This is due to a combination of factors: the spectacular growth of China, the economic advantages deriving from large pools of well-educated and highly skilled workers and managers, increasing intra-regional and global trade, growing internal demand, realization of the benefits of several years of economic reforms, and marked shifts in the locus of production of goods and services from industrialized nations to emerging economies of the region due to relentless cost-cutting pressures in industrialized nations.

Many parts of Asia possess the fundamentals—talented human capital, strong educational traditions, and relative openness to trade, ideas, and investment—required to excel in the knowledge-driven global economy of the early twenty-first century. Translating these into long-term success requires sustained application of good economic policy (especially further efforts to develop and nurture efficient and credible domestic financial markets and to strengthen corporate governance), and a well-functioning global financial and trading system.

In the near term, two central questions will face investors in the region: One, how will economies and markets react to a firming of interest rates in the United States or a sharp fall in the dollar against Asian currencies? Two, will China's growth slow, and if so, what will the impact be on other Asian nations? On the first question, most Asian economies should be able to withstand moderately higher U.S. interest rates—as they did in the late 1980s—or a gradual decline in the dollar, as long as they themselves can maintain accommodative financial conditions and sustain steady increases in domestic demand. However, a very sharp decline in U.S. growth, or in the dollar, would have serious implications for the region because of Asia's continued heavy dependence on exports to North America.

A significant slowdown in China, while highly unlikely in the near term, could become a problem for the rest of the region if it were to occur. Many of China's neighbors have substantially boosted sales to that nation in recent years; a sharp deceleration of China's growth would cause commodity prices to tumble and other sales (particularly of intermediate goods) to weaken—with ripple effects in equity and commodity markets throughout the region. However, most countries could ride out a modest slowdown in China's growth due to the cushion afforded them by large currency reserves, flexible exchange-rate regimes that would permit a lowering of currency values to boost exports to other markets, the inroads many of their products have made in the U.S. and EU markets, the ability to run larger budget deficits for a time, and a more diverse product and export mix than a decade ago.

The Growing Role of Domestic Demand in Driving Asian Growth

After years of significant dependence on export-led growth, one of the broadest changes taking place in Asia is the prospect of a stronger role for domestic and regional demand in driving growth. If this process is to be sustained, it requires more investment in industries not related to exports and a broader base of consumer demand. Progress in such areas will be particularly important to sustaining the stability and improving the balance of the world economy in coming years. Most Asian economies continue to be significant exporters of a wide range of goods and services to the United States and the EU—and final demand in those economies remains critical to the economic health of the Asia region. A recent International Monetary Fund report pointed out that of all exports from emerging Asia, 81 percent go directly or indirectly (in the form of exports to other Asian nations for assembly into final products shipped outside the region) to non-Asian industrialized countries.

For many emerging nations of the region, however, increasingly strong growth in capital investment and in consumer demand, supported by increases in domestic credit creation and accommodative monetary and fiscal policies, hold out the prospect of reducing their relative dependence on exports as a source of overall economic activity. The trend is just beginning, but if reinforced and sustained, it will help to rebalance global growth and reduce international trade disparities, particularly with the United States but also with the European Union.

Key to whether this process can be sustained is China. Last year it applied measures to cool down the economy in order to reduce inflationary pressures. This has moderated domestic demand growth in recent quarters and made overall growth relatively more dependent on exports; but domestic demand is still likely to remain robust. China is likely to continue to experience annual trade deficits with most of the rest of Asia, but rising surpluses with the United States or the EU. Robust domestic demand in the future will depend in part on continued redeployment of China's high domestic savings and from inefficient state enterprises and into more productive investments, broader-based growth in rural China, and a currency adjustment.

Domestic demand is strengthening elsewhere in Asia. In many countries, it is being led by a resurgence of investment spending. Capital spending has been sluggish in much of Asia for some time, largely because the region has been recovering from, first, a period of dramatic overinvestment in the mid-1990s, then from the 1997 financial crisis, and then from the effects of severe acute respiratory syndrome (SARS). It also bore the effects of the post-bubble weakness in demand for high-technology products in industrialized nations.

As the aftereffects of the 1997 crisis wear off and technology spending in the G7 nations strengthens, investment in the emerging economies of the region, particularly in industries geared to high-technology exports, is gaining momentum. That, in turn, is boosting employment and private consumption. The improved condition of banks in the region, due to reform and restructuring, together with their greater emphasis on consumer-oriented financial products, such as mortgages and credit cards (which banks advanced in recent years to reduce dependence on sluggish export financing business during the global downturn), also has boosted prospects for stronger Asian domestic consumption.

There is also a new spirit of intra-regional cooperation, and a strong push to put in place more formal institutions to facilitate that cooperation. The 2001 Chiang Mai Initiative produced agreement on commitments of $35 billion for bilateral currency swap arrangements, although improving payments balances have virtually eliminated the need to draw on these now. As intra-regional trade continues to grow, there is increasing focus on broader currency cooperation—and ways to avoid excessive exposure to the dollar. One goal is to make a larger share of investment in local assets. The Asian Bond Fund Initiative (ABF 2)—announced by the central banks of East Asia, Australia, and New Zealand—would create an Asian Fund of Funds, which would invest in bond funds in individual East Asian countries. These countries also agreed to create a Pan-Asian Bond Fund Index, which would invest directly in local currency bonds in the region, enabling international investors to buy a diversified portfolio of regional debt.

China

With increased financial, trade, and investment linkages to its neighbors, any analysis of the economic future of Asia needs to focus significantly on the economic outlook in the Middle Kingdom—an ancient political appellation now also appropriate for economic reasons. Less than 200 years ago China's gross domestic product (GDP) exceeded that of any other nation. (Second was the Indian subcontinent—the area later, and temporarily, known as British India.) The Chinese remember their history. Many see their recent surge in growth as a comeback rather than a surprising new phenomenon—which is how most Westerners see it.

China is at the center of the unfolding and impressive Asian growth and trade story. In the past two decades, China has achieved remarkable increases in output, job creation, investment, exports, and living standards—and in so doing also has given an enormous economic boost to the rest of Asia and many economies outside of the region as well. It is now a world-class economic power and a major factor in the international economy. But China also faces important policy challenges, which include high unemployment in some regions, formidable banking problems, many loss-making state enterprises, energy shortages, rural poverty, social security deficiencies, and inadequate food output due in part to reduction in farmland.

Until recently, the task of overcoming deflation occupied Chinese macroeconomic policymakers. Expansion of bank credit and monetary growth in 2002 was part of an all-out effort to combat this affliction. In 2003, however, prices began to rise—good news for China and for many other Asia nations that had been concerned about China exporting deflation. But the prospect of too much inflation in 2004 posed serious problems—and threatened to create enormous imbalances. Chinese officials had worked hard to combat rapidly rising prices in the past decade (double-digit inflation in 1993 and 1994). In 2004 they again turned their attention to coping with a resurgence of inflation. Inflation in the early 1990s (a product of the torrid pace of investment then) was curbed when the government, utilizing a combination of "administrative guidance" and severe macroeconomic (particularly credit) tightening, brought prices down dramatically. However, the country subsequently experienced a troublesome run of deflation and weak growth.

China's 2004 increase in inflation stemmed in part from sharply higher food prices (due to a poor harvest and shrinkage of farmland, as more of it was turned into commercial real estate) and sharply rising fuel prices. It could also be partly attributed (as in the early 1990s) to the nation's investment boom. China experienced an explosion of construction projects, mostly sponsored by local authorities, in sectors such as steel, cement, aluminum, commercial real estate, and infrastructure. That was manifest in price increases in real estate, construction materials, and commodities. Companies also were buying up materials and commodities for inventory (as a hedge against further price increases) at a torrid pace, thus exacerbating the price increases.

Strong investment (currently at near 40 percent of GDP) will continue to be needed to create new jobs in order to absorb the millions of workers laid off by the closing or downsizing of large state-owned factories (forty million jobs lost over the past five years) along with those coming to China's cities in the industrial east from rural areas in search of jobs (about ten million annually) and new entrants into the labor force (also about ten million annually). But too much investment in certain sectors created bottlenecks, thus triggering sharply rising prices in some areas of the economy. Moreover, a number of new projects were discovered to be unsound economically (many projects produce very low returns), of low quality, and placing strains on an already overtaxed energy system.

Beijing's Balancing Act

The central government's decision to set the 2004 growth target at a modest (for China) 7 percent was part of a broader effort to contain inflation and produce a more sustainable rate of growth. Chinese authorities did not actually want to drive growth down to that level. Announcing this figure was a way to signal China's provincial and local governments (which account for about two-thirds of all investment projects in China, more than five times that of the central government) to curtail project development—which in the first quarter of 2004 grew at a rate over 40 percent higher than the comparable period in 2003.

However, the ability of Beijing to control the provinces and cities, and curb project development there, has slipped over time as the result of years of reform that decentralized economic decision-making and resource allocation. Often local governments have sufficient power to persuade local banks to fund their pet projects, whether the banks want to or not. This is not to say that the central government cannot accomplish the task, but it is more complicated than in the past.

In one fundamental sense, the problem of the Chinese economy in 2004 was not one of classic overheating. That normally is characterized by tight labor markets and rapidly rising wages. China has high unemployment, and wages remain low in much of the country. In many areas of manufacturing there was overcapacity. China's problem was capacity/supply constraints in some sectors that raised prices in those sectors.

Beijing periodically has to engage in a delicate balancing act—attempting to cool overheated sectors while increasing investment in, and encouraging more private investment in, areas and sectors that are high national priorities (such as rural China and power generation), and where bottlenecks need to be reduced (e.g., transportation, coal, and oil), and maintaining the overall growth momentum at a rate sufficient to generate increased employment. The government tightened credit and raised interest rates in 2004, and could do so again if inflation picks up. But it will be reluctant to risk a "hard landing." Unemployment is already a concern—and Beijing wants to prevent its rising.

Authorities also are concerned that higher interest rates in the future could draw in even greater capital inflows from abroad. Rising capital inflows—some of them speculating on an increase in the value of the currency—are a major reason that foreign exchange reserves increased so rapidly in 2004 (and early 2005), making management of the money supply and credit more difficult—although the People's Bank of China so far has managed to moderate growth in the money supply very effectively. Rate increases by the Federal Reserve enable the People's Bank of China to raise rates with less chance of such inflows from dollars into renminbi (RMB).

In 2004, Beijing used a combination of tighter macroeconomic and administrative policy. It twice raised bank reserve requirements to slow lending and sent officials to several provinces to audit bank loans and investments with the goal of limiting new lending in overheated sectors, and disciplining officials who engaged in unsound lending practices. Among its toughest measures was the order given in late April to halt all new loans temporarily—giving the government time to assess the problem and further reign in what many authorities believed to be reckless lending practices. Other measures included orders to local authorities to require project sponsors to have more cash before undertaking a project, instead of borrowing a large portion of the cost, and tighter supervision of joint-stock banks, which engaged in especially exuberant lending growth—in some cases 50–60 percent increases.

Authorities in Beijing also raised interest rates and gave banks the option of charging much higher premiums to riskier borrowers or to projects in overheated sectors, while providing the lower benchmark rate to less risky borrowers and projects the government deemed desirable (such as electric power and railways). This illustrated a key aspect of the government's strategy to reorient investment—creating strong pressures to curtail some types of projects but redirecting capital to others that it wanted to go forward. For the most part the strategy employed in 2004 was a success; real fixed asset investment growth cooled from roughly 25 percent to 17 percent, and shifted away from overheated sectors such as steel and cement. Output in priority sectors such as food and electricity increased significantly. Inflation fell. All this had been done while GDP growth remained around 9 percent.

A large increase in unemployment in the ongoing transition from an economy shifting from heavy dependence on agriculture to greater dependence on manufacturing and services and at the same time from one in which state-owned enterprises (SOEs) are the main employer to one in which private enterprises ultimately play that role could be destabilizing. Beijing will do what is necessary to avoid high levels of unemployment and to ensure that the transitions are as smooth as possible. China cannot afford to have the investment sector cool too much, because a very large portion of China's growth and jobs comes from new investment.

A rapid deceleration of investment growth could slow the economy and the jobs market more than the government wants. That is why the rebalancing of investment described above is so important. China has a better chance of avoiding volatility now than in the past because its officials have gained valuable experience. They also recognized the problem of inflation earlier in the current cycle than in the early 1990s and acted quickly; overall economic conditions are better than at that time.

Concern over the viability of SOEs—and the government's desire to avoid major increases in layoffs—has caused the government to direct most available capital there. That in turn has limited the access of private enterprises to bank credit or other types of government support. Private companies have the potential to create significant numbers of new jobs when they have additional access to financial resources. Nevertheless, for the moment a large portion of all bank lending goes to SOEs. Improvements in domestic capital markets are needed to boost private sector access to equity and debt funding.

The Prospects for Currency Revaluation

Concerns about inflationary pressures, the reluctance of the central government to engage in a dramatically tighter monetary policy or a sharp contraction in fiscal policy to hold down inflation, and large inflows of speculative funds increased talk of a currency revaluation or float beginning in 2004. Such talk, and pressures from foreign governments, picked up momentum in 2005. A higher value for the RMB would dampen inflationary pressures by reducing the price of imports. The difficulties in managing massive reserve accumulation and the prospect that Chinese banks (like many in Asia) will lose their appetite for large amounts of "sterilization bonds" (bonds issued to sop up local currency created when the government intervenes in the foreign exchange market to buy up dollars) are related reasons for allowing the currency to rise in value.

Chinese authorities, however, have indicated that they do not believe that it is appropriate to use currency appreciation as a short-term policy instrument for curtailing inflation. For several years they have seen a stable RMB as a key element in their overall policy effort to sustain investment inflows and promote a stable financial environment. They underscore that their goal is to engineer steady growth and job creation. They point out that a significantly higher RMB could work in the opposite direction by slowing growth too much and adding to unemployment. Moreover, they believe it could trigger volatility among Asian currencies, thereby disrupting tightly knit supplier relationships in the region. They further argue that a small increase would do little to stop speculation or inflation—and could cause markets to assume that further increases will be forthcoming, thus increasing speculative inflows. In any case, China has repeatedly said that its goal is not so much to revalue its currency as to reform the nation's exchange rate regime, which could include pegging the RMB to a basket of currencies and/or widening the band within which the RMB fluctuates; a revaluation of the currency vis-à-vis the U.S. dollar would likely take place in that context.

The emergence of an informal RMB zone in East Asia attests to just how important China's trade with the region is and underscores the significance of the RMB. While many East Asian and Southeast Asian countries would benefit from a unilateral revaluation of their currencies to help manage money supply growth, several governments fear that if they make a move on their own, without an appreciation of the RMB, or act too early, they would risk losing market share, particularly vis-à-vis China.

A revaluation of the RMB, or a controlled float in a wider band, would enable (or encourage) other emerging economies in the region to raise the value of their currencies. In part this would result from markets responding to a Chinese revaluation or upward float by putting upward pressure on other currencies, and in part from the willingness of their governments to allow their respective currencies to rise, knowing that their economies would not suffer a fall in competitiveness vis-à-vis China as the result. The collective outcome could be a revaluation of regional currencies against the dollar and the Euro. Because Asian currencies together constitute nearly 40 percent of the dollar's trade-weighted exchange rate, this would be a step toward reducing the large U.S. trade deficit with the region. However, it would not cure the overall U.S. trade deficit, which is largely the result of a high level of U.S. consumption and a low level of domestic savings relative to domestic investment. Moreover, there is very little overlap between the products China and most other mainland Asian countries produce and those that the United States produces, suggesting that the result of such revaluation might be higher prices for Asian imports but not a major reduction in U.S. import volumes.

China's Competitive Strength

China's enormous demand for commodities and semiprocessed goods has contributed to economic recovery in much of Asia—and to considerably higher prices for many Asian raw material and industrial exporters. Many firms in Asia also have had to face stiffer competition as China's comparative advantage in light industry and electronics assembly has eroded their exports. But many others have come to see China as an enormous marketing opportunity. They are focusing on China not as a competitive threat, but as a huge and growing market for primary, intermediate, and finished goods. A quick look at a few statistics explains why: In recent years, China has consumed 25–30 percent of world steel output, 30 percent of world coal output, 50 percent of world cement production, and 10–20 percent of world electricity. And many Asian companies—like those elsewhere—have increased their global competitiveness by investing in China to take advantage of the increasing skills of its large, low-cost workforce and have been successful in boosting sales in China and exports from China.

It would be a mistake to see China as competitive just because of its cheap labor. Increasingly it will be skilled labor and high-quality engineers and scientists—working in well-funded research/development/manufacturing hubs—that propel growth. Investors interested in China should focus increasingly both on its growing domestic demand and on its ability to produce higher quality, high value-added goods, rather than simply its ability to make goods cheaply. Nonetheless, low-cost labor will continue to be the centerpiece of China's exports and growth for a long time. Manufacturers are increasingly locating plants in low-cost areas in the Chinese west, rather than more expensive eastern and southern coastal areas. Government tax and infrastructure improvements in western regions are facilitating this process.

The "Five Balanced Aspects"

The Chinese authorities' emphasis on the development of western China must also be seen in the context of concerns over rural discontent resulting from disgruntled farmers and migrant workers. Authorities recognize that agricultural regions have not benefited as much from national prosperity as urban eastern and southern regions—especially cities along China's two thriving river basins. Correcting rural-urban and western-eastern imbalances are two parts of the policy of the "five balanced aspects" decided on at the March 2004 annual meeting of the National People's Congress (NPC).

The NPC committed the central government to pay more attention to western provinces and to the development of China's rural peoples; investment incentives have been enhanced to shift more resources to these regions. Other parts of the "five balanced aspects" are rebalancing:

- between economic and social development, in favor of the latter;
- between man and nature, in favor of the latter (in response to grassroots environmental pressures); and
- between the domestic economy and the foreign sector (with the goal of strengthening the former and relying less on the latter).

The Demand for Energy

One area of the economy that merits special attention is electric power. Power shortages have been a common problem in China for years. The much-publicized shortages of 2002 were largely the result of the fact that the government tightly controlled the price of electricity, but not the price of coal. Regulated utilities were limited in the amount they could pay for coal, so some coal producers sold abroad instead, at a market price. Serious energy shortages are likely to present a medium-term problem; rolling blackouts and brownouts disrupt production and concern government officials.

China has launched a major effort to improve the efficiency of energy usage. A significant portion of China's energy is wasted through inefficient practices. It is also seeking to increase the supply and security of energy by investing in oil and gas resources throughout Asia, the Americas, and the Middle East; strengthening political and economic cooperation with oil- and natural gas-exporting nations; boosting domestic oil, gas, and coal production; and constructing pipelines from oil and gas fields in neighboring countries.

China has been very successful in developing energy opportunities in Asia and the Middle East. However, efforts to boost domestic production have encountered difficulties; new, low-cost fields have been hard to find. Efforts are being made to introduce more foreign technology and expertise and develop alternative sources. Large new offshore areas have been offered to foreign companies for development. Coal is China's single largest source of energy, and almost all of it now is in government hands. Privatizing some facilities and expanding the use of new technology to ensure that new coal projects conform to higher environmental standards will open important opportunities for domestic and foreign firms and joint ventures.

Russia earlier had appeared to favor a pipeline from its Siberian oil fields into northern China. It now appears to favor a pipeline to a port in the Russian Far East, to ship to Japan and elsewhere (including China). That strategy, the Russians argue, would provide a broader range of marketing options and opportunities compared to a pipeline exclusively dedicated to China. However, China and Kazakhstan are building a large pipeline—3,000 kilometers long—to transport Kazakh oil into western China.

By the end of this decade, China is likely to become the world's largest importer of natural gas and an enormous importer of oil. Refining and distribution are opening to foreigners. China's own companies, many of which have floated stock in international markets, are already very competitive in this area. These companies have a strong domestic lead and a lot of expertise, but foreign companies will have significant opportunities in coming years. As an indication of how international China's energy companies have become, in 2004 Sinopec was awarded a large concession in Saudi Arabia to develop natural gas, and in 2003 it was awarded a concession to develop oil fields in Iran. Led by the desire to secure energy and raw material production around the world, China is likely to become one of the world's largest foreign investors during this decade.

Cooling-Off Period

In the period ahead, China's economic outlook will turn on how successful the government is in sustaining steady growth in jobs and overall economic activity without triggering higher inflation, avoiding an overheated boom followed by a deflationary bust. Authorities in Beijing are likely to continue to rely on administrative measures along with macro-policy—taking incremental rather than dramatic steps—to cool sectors that have experienced overheating or overinvestment and to direct more capital to relieve bottleneck sectors. Success in "rebalancing" and sustaining growth, with more investment going to agricultural and western China, relative to the eastern, industrialized parts, will be important. That will take some pressure off the eastern jobs market to generate rapid employment growth and should help to reduce rural discontent.

Cooling down the housing sector presents another challenge, especially in places such as Beijing and Shanghai. The government already has imposed a 5.5 percent flip tax on sellers to discourage real estate speculation. But cooling must be done carefully lest banks be saddled with more bad loans if the housing bubble bursts.

In the medium-term the economy will also benefit from strong overall consumer demand, as increased earning power enables managers and skilled workers to buy home electronics products, cars, and other higher-end products. All told, while China's investment boom is likely to moderate and speculative pressures in the property markets are likely to dampen, new drivers of domestic demand, along with continued strong export competitiveness, are likely to sustain a healthy rate of economic activity with growth rates of more than 8 percent in 2005 and 2006. The high quality of China's economic leadership, together with its thoughtful approach to policy-making, is another of the country's long-term assets and should give added confidence to investors. By 2050 China's GDP could rise above $40 trillion (in 2003 U.S. dollars), making it the world's largest economy—again!

Recovery Around the Region

The domestic credit cycle in much of the rest of Asia has turned markedly positive—improving chances of a sustainable recovery in the region outside of China. Bank lending to households and corporations has picked up. This is evidence of continued recuperation from the post-1997–1998 financial trauma. Immediately after that, the primary goal of "crisis economies" was to stabilize the balance of payments of crisis affected economies (a goal made easier by the introduction of more flexible currency regimes) and their banking systems. Problems in both areas were the principal causes of the crisis. During this first phase of the recovery, monetary policy remained tight, banks were recapitalized, and domestic demand was restrained. Combined, these led to a sharp improvement in payments balances.

After the stabilization of payments balances and improvement of currency reserves and bank financial conditions, capital began to flow back to the region. Some went to buy depressed assets, including the billions in nonperforming loans (NPLs) that banks were selling. Over the past five years, around $1 trillion of Asian NPLs have been disposed of. Weaker currencies following the 1997 currency collapse led to improved trade competitiveness, boosting exports, restraining imports, and further improving trade balances. Currencies began to strengthen in time as the result of capital inflow and trade improvements. Interest rates eased as capital returned, banks became stronger, and default risks (which kept spreads high during and for a while after the crisis) diminished. In turn, banks engaged in more consumer lending, and businesses and consumers enjoyed the benefits of stronger growth from exports. Corporations continued to reduce their debt, which had rendered them so vulnerable to higher rates and a drop in domestic demand during the crisis.

Now countries are realizing the fruits of their efforts—especially the completion of "supply-side" adjustment in the form of stronger corporate balance sheets and the improved lending capacities of the now stronger banks. They are starting to see a pickup in corporate capital spending, which in turn is driving increases in disposable income and greater consumption.

Malaysia

Recovery is very much in evidence in Malaysia. Capital spending is rising after a long period of weakness, and investors are switching from less risky assets such as cash and bonds to higher risk assets such as stock. Consumers have benefited from increased availability of domestic credit, a rise in rural income from higher commodity prices, and the spillover benefits from the capital spending recovery. Malaysian consumers also have benefitted as interest rates have eased and wages have picked up due to higher exports of information technology (IT) products.

Indonesia

In Indonesia, the accelerated disposal of nonperforming loans, improvement in banking and the corporate sector, the fall in interest rates, and prudent fiscal policy have boosted capital investment—now accelerating at 8 percent annually. Raw material and energy exports to China have also given the economy a boost. Domestic demand is growing in the 5–6 percent range.

Taiwan

In Taiwan, household lending firmed over several quarters, and corporate lending is kicking in. Taiwan has benefited from increased economic activity, due in significant measure to trade with mainland China. The IT sector has been picking up, based especially on rising sales to the American market. Domestic lending for construction is boosting that sector and more infrastructure investment is on the way. Domestic fixed investment should average around 10 percent this year and next, and domestic demand around 5 percent. Political tensions with the PRC—and their risk of disrupting the island's growing commerce with the mainland—remain on the minds of investors.

South Korea

South Korea was faced with an economic slowdown and job loss during much of 2003 due to a sharp fall in domestic demand, significant bankruptcies, and large numbers of credit card delinquencies. Subsequently, there was a fall in household borrowing as the credit card bust was unwound. The country experienced a decline in private consumption in 2003 as well as in corporate borrowing. Both have reversed in 2004, the former due to an overall pickup in exports, and the latter because of strengthened capital spending. This is the result of a combination of factors: low interest rates, improving business sentiment, a general pickup in the IT sector worldwide, the desire of business to invest in the next generation of technology, and continued government deregulation. South Korea's world-class IT infrastructure and large pool of highly educated managers and workers are also attractions to investors.

The South Korean government frontloaded stimulus in 2004. That, plus low interest rates, was expected to lead to a comeback in employment and household consumption. Nonetheless, domestic consumption has remained sluggish due to higher oil prices and the legacy of credit card delinquencies, and recent export momentum has weakened, because of a stronger won and cooling foreign demand for technology, producing a modest growth outlook of 4 percent for 2005, rising to 5 percent in 2006 contingent on a resurgence of capital spending.

The Beginnings of an Expansion

All told, the domestic credit expansion cycle in the Asian region probably has a way to go—as does the revival of domestic capital spending. Many countries continue to pursue an export growth model, as evidenced by their desire to sus tain currencies at relatively low levels against the U.S. dollar. Credit expansion, however, is likely to continue because in much of the region firms are poised for higher investment; they have just emerged from a multi-year effort to reduce their debt and now have a great deal of cash on their balance sheets with which to make capital investments. They are likely to do so if they foresee sustained growth in global or regional demand. Willingness to take risk has also grown, which will reinforce their desire to make new investments, and to incur new debt if they see stronger profit opportunities. This same increased appetite for risk is leading banks to move away from safe government bonds in favor of additional private sector lending. And unlike the 1990s, there is considerable liquidity in most of the region's economies, so additional domestic lending can take place without recourse to offshore funding—the great magnitudes of which got banks and economies into trouble in the late 1990s.

This desire of banks and investors to take on more risk and obtain higher returns complicates currency intervention policy in much of the region. That in turn strengthens the argument for emerging economies of the region to allow their currencies to rise in 2005, especially (for reasons mentioned above) if China takes the lead. East Asian nations have been building up enormous currency reserves since the end of the crisis. But the pace has accelerated rapidly over the past three years because governments have sought to resist the fall of the dollar against their currencies through large and sustained intervention in the currency markets.

Banks earlier were concerned primarily about deflation and had a high aversion to risk, so government intervention could be "sterilized" relatively easily. That is, banks were generally willing to buy the "sterilization bonds" (as in China) issued by their nations' treasuries to sop up the extra local currency generated by intervention. (Buying dollars by selling, say, *won* generates excess won liquidity that must be "sterilized" by the selling of won bonds, lest domestic money supply rise too rapidly.) Now, banks are seeing new lending opportunities in the private sector, so they are more reluctant to buy sterilization bonds—and many are reluctant to roll over the ones they own. This puts added pressure on governments to either raise the interest rate they pay on these bonds (which would suck in more money) or allow currencies to rise in order to reduce the need for massive intervention—with the alternative being a continued, unwanted expansion of domestic money supply triggering a credit boom that could lead to inflation later.

Challenges for Southeast Asia

Southeast Asia faces a particular set of challenges. The nations of ASEAN (the Association of South-East Asian Nations, which includes Brunei, Cambodia, Indonesia, Laos, Malaysia, Myanmar, the Philippines, Singapore, Thailand, and Vietnam), a once fast-growing region, have struggled to regain the formidable economic strength they showed before the 1997 financial crisis. Foreign direct investment has been a lot weaker than before; so has their rate of economic growth. Wages have also crept up. And in most manufactured products China has a decided competitive advantage. One of the area's expected potential strengths—its free trade area established in 1994—has proved one of the greatest disappointments; intra-regional trade has fallen as a portion of overall trade.

The next few years provide an opportunity for the region to demonstrate more robust economic performance. It possesses great strengths: a skilled work force, strong education systems, an abundance of natural resources, for which China will be a growing market, and a wide range of high-technology industries centered on consumer electronics and information technology. Yet it will fall short of realizing that potential unless it creates a genuinely integrated market instead of the still fragmented one that exists today—which discourages foreign investment and leads to an inefficient allocation of resources in the region. Lack of common product standards, a host of regulations that protect local industries, and favoritism that enables unproductive companies to remain in business produce undesirable results.

The decision in 2003 to create an ASEAN Economic Community provides a renewed opportunity to implement reforms, focusing on reducing trade barriers and increasing market efficiencies that increase productivity. With China and now India surging ahead, and providing magnets for foreign investment, the pressure will be on ASEAN nations to strengthen competitiveness and attractiveness to direct investors. The region is likely to do better in coming years than in the recent past, but whether or not it can regain its former vigor can make the difference between rapidly rising living standards and only modest improvements.

Japan

If increased capital spending and increased trade within East Asia are to be features of the growth story in Asia over coming years, Japan is showing signs that it will be a more positive participant in that story than in the recent past.

After being a drag in the region for much of the decade, Japan is showing signs of becoming a generator of growth. However, the process has been highly uneven. Roughly three-fourths of Japan's remarkable 7 percent annualized growth rate in the fourth quarter of 2003 came from domestic private demand—and only one-quarter from foreign demand. The same was true for the first quarter of 2004, when Japan achieved a 5.6 percent growth rate. Capital spending was a big reason for Japan's increased domestic economic activity—up 30 percent in the fourth quarter of 2003 and 10 percent in the first quarter of 2004. Consumption rose at around 4 percent. However, in the middle of 2004 the country slipped back into recession, as consumer spending slowed and exports cooled. It now, once again, is growing, experiencing a roughly 2 percent rate of increase in consumption and 5 percent for capex.

In 2004, the unemployment rate dropped for the first time in thirteen years. Wages have also begun to improve, after declining on an annual basis for several years. And job prospects are now improving as well; consumer spending is also picking up. Increased capital spending by large companies is helping to improve the outlook for small and medium-sized companies, many of which depend on large companies for sales.

Much of Japan's new investment is now aimed at meeting growing foreign demand, particularly from China. East Asia accounts for almost half of Japan's exports. Since mid-July 2001, Japan's exports to China have grown dramatically; early in 2004 Japan had a trade surplus with China, its first in over a decade. However, net exports, as noted earlier, are not the dominant factor in Japan's growth. Exports account for only 12 percent of Japan's GDP—not all that different from the United States. Sustainability of strong growth depends on continued expansion of domestic demand. The recent pickup in the nonmanufacturing sector (such as real estate and retailing) and in telecommunications, information services, and corporate services have improved prospects of broad and sustainable increases in economic activity.

In light of the current outsourcing controversy in the United States, it is useful to reflect on Japan's circumstances. Growing numbers of Japanese companies have engaged in plant construction in China to take greater advantage of the growing China market. Small and medium-sized Japanese companies have established themselves as suppliers of automotive equipment and electronic components to larger Japanese companies—and other companies as well—in China and throughout Asia. And more and more Japanese companies source components in China and elsewhere in the region for incorporation into final products assembled in Japan—thus lowering their cost structures.

Core consumer prices, which the central bank uses to measure inflation, have recently stabilized, with projections for a pickup in coming quarters. If Japan can grow at 1–2 percent annually for the next couple of years, the output gap—which has been the primary reason for Japan's structural deflation—will disappear.

Young people have suffered most from recent years of employer caution; unemployment among men under 24 stands at around 10 percent. Until wage and labor conditions remain positive for a few quarters, it will be difficult to predict strong, self-sustaining growth and a better outlook for this portion of the population. But prospects appear good for both if hiring continues to pick up and the central bank continues to pump money into the economy and keep interest rates near zero—which it has pledged to do until deflation is overcome.

Several factors will affect the outlook for Japan. One will be the future course of the exchange value of the yen. Big increases in the currency's value will weaken growth prospects—but such pressure has abated recently. Sharp upward pressure in the future likely will be resisted by government intervention in currency markets. Moreover, as Japanese companies become more competitive and domestic demand grows, many of them can make profits at a higher yen valuation.

Growth in the United States, China, and elsewhere offset most of the trade effects of the yen's rise in 2004 and early 2005 by increasing the volume of Japanese sales overseas. A rise in the value of the RMB and other non-Japan Asian currencies against the U.S. dollar would also reduce the negative impact of the higher yen against the dollar assuming the yen does not also rise again; Japanese products would then become more competitive against those of these countries.

A second factor affecting Japan's outlook hangs on whether improved domestic economic conditions lead to relaxation of restructuring pressures. Japanese corporations and banks have made considerable progress in reducing leverage, reducing bad loans, and improving business profitability. If, as some argue, the 1990s was a "lost decade" for the Japanese economy in terms of growth, it was not for Japanese banks and corporations, or government policy; important reforms took place. These reforms are paying off. Deleveraging and stronger balance sheets are evident throughout corporate Japan. Many companies have sold off or closed unprofitable businesses, thus narrowing the large excess of supply over demand that was the key factor driving deflation. Many have strengthened global supply and production networks—largely through additional production of final products elsewhere in Asia, and sourcing in lower-wage economies in Asia of components for incorporation in higher-value products at home.

The restructuring process has a way to go if it is to purge the economy of the remaining bad loans that are the legacy of the excess lending and borrowing "bubble" of the 1980s. Continued restructuring is needed to put the financial sector and many overextended companies in a stronger financial position—and to further reduce the still large number of inefficient suppliers being sustained by the banking system. A continued sense of urgency and commitment is needed to achieve further reductions in nonperforming loans.

Nonetheless, the Financial Services Agency (FSA) has documented dramatic improvements in the NPL picture—showing that the amount of NPL disposals is decreasing, as is the need for setting aside additional funds for loan-loss provisions. The amount of bad loans held by the biggest banks declined by 72 percent from 2002 to 2005; and the bad loan ratio dropped from 8.4 percent to 2.9 percent. The FSA also reported that loan quality of significant numbers of banks was revised upwards. At the same time, to keep the heat on, the FSA is monitoring a few target banks and a few troubled borrowers—thus attempting to ensure that recent improvements do not lead to complacency.

India

A relatively new entrant in the global competitive race is India. Given the enormous talent of its people, economists had been asking why India had not been a bigger player in the global economy. For many years it was derided for its low "Hindu rate of growth." But in the last decade and a half, major changes have taken place. Technology has boomed, tariffs were cut, the oil and gas sectors were opened to foreign participation, the banking sector became more open, and many formerly state-owned enterprises were successfully privatized.

The surprising victory of the coalition led by the Congress Party over the National Democratic Alliance led by the BJP (Bharatiya Janta Party) led some observers to ask whether these changes would be sustained. The economy had grown rapidly, and many in India's upper and middle classes have done quite well. However, many voters in India's vibrant democracy, in which large numbers of people (a larger percentage than in the United States) turn out to vote, evidentially did not believe this growth helped the struggling rural sector—in which the vast number of Indians live—or the urban unemployed seeking jobs on the streets of the country's teeming cities.

The election results raised the question in the minds of some investors as to the extent to which increased emphasis on distributive justice would slow growth, or whether demands for large subsidies and more spending in rural areas will reverse the progress that had been made in reducing the country's still large budget deficit. Communist supporters of the incoming government had been strong proponents of big subsidies on such staples as food, fertilizer, and electricity, and Congress had pledged to "attain and sustain" an 8–10 percent growth rate, which, some felt, would require a big budget deficit that would crowd out private investment.

This surprise election result should serve as an important reminder to investors in emerging economies that in most of these countries the rural and urban poor constitute a large portion of the population. Policies that do not address their needs will come under public criticism. Governments and policies that might look good to foreign investors but unattractive—or unhelpful—to large numbers of the discontented masses within a country might lead to policy changes, government changes, or both.

The coalition that governs India now includes members of the country's pro-labor, pro-rural parties, many of whom have criticized deregulation and privatization for their effects on the poor. (The Communist Party–Marxist, which supports the new government and holds the third largest number of states in the new parliament, remains outside of the formal coalition.) One of the first moves of the new government was to announce that planned privatizations of the country's big oil refineries would be cancelled.

However, the concerns that some in the financial community originally expressed about the new government proved greatly misplaced. Manmohan Singh, the prime minister, is an outstanding and highly respected economic and political leader. He began the country's reforms in 1991, when as finance minister he dismantled the complex system of government licensing in most sectors, cut tariffs on large numbers of imported goods, and opened important sectors of the economy to foreign investors. When elected, he assembled an impressive economic team—one of the best and most respected in the world. Moreover, in the states of West Bengal and Kerala, where the Communist party has governed, its leaders have supported many reforms; in West Bengal, for example, the government has outlawed strikes against information technology-enabled services. So the prospect of a reversal of the reform process is highly unlikely. But modifications of the previous government's policies have taken place—with more emphasis on addressing the problems of India's rural poor through rural credits and infrastructure projects. Success in improving rural and urban living standards would provide a solid base—economically and politically—for continued economic growth in India and for a broader consumer base.

Whereas most of Asia began its economic boom based on industrial expansion, India's boom has been primarily services-led. Services have grown by roughly 6 percent annually over the past ten years. The services sector, which accounted for 35 percent of GDP in the early 1960s, is around 56 percent of GDP today. Manufacturing has risen from 15 percent to 22 percent of GDP during that period. Information technology (despite all the international headlines) is only a modest 3 percent of India's GDP. But it has tripled as a portion of GDP over the past five years and, given the right policy environment, could triple over the next five. Its success is based on its ability to draw on India's large pool of skilled professionals and the sizable portion of the population fluent in English. It has thrived in an IT environment freed of heavy regulation. Deregulation of communications has facilitated rapid cost cutting and vigorous competition, much as it did in the United States two decades ago.

Since the mid-1990s, service exports have been the driving force behind the doubling of Indian exports. Business process outsourcing and information technology sales are important factors in this growth. Yet India still accounts for only 1 percent of world exports. India's aim is to increase exports by 12 percent per year through 2007. Stronger export growth has raised the level of foreign exchange reserves dramatically, which in turn enabled the country to cut peak duties. Because lower duties will lower the import costs of industrial inputs, including metals, they also will give a boost to exports such as automobiles. Other measures, such as the simplification of customs procedures, will help users of imported materials and components as well.

India and China (although different in many ways) are the most competitive of the emerging Asian economies. Therefore, a comparison of the two countries and their economic characteristics is useful. But any comparison must take into account a key fact: China's drive toward liberalization and reform began in 1978, when Deng Xiao Peng initiated the policy of openness to foreign investment and trade. India's reforms began over a decade later, when the then finance minister, Manmohan Singh, introduced a series of pro-competition, pro-private enterprise reforms. Since then, India has reduced or eliminated many of the impediments to economic activity that collectively had come to be known as the "License Raj"—a bewildering maze of licensing requirements, price controls, capital allocation regulations, and so forth. However, the nature of the government process in a vibrant democracy like India, where coalitions can change quickly (witness the May 2004 elections), often in the past made it harder to implement reforms in a smooth fashion. The direction of India's reforms today appears to enjoy a broad consensus.

Around twenty years ago India's per capita GDP was about the same as that of China—$275. In 2004 it had been around $500, still less than half of China's, at over $1,000. India's population growth rate has been falling, from 2.1 percent in 1990 to 1.6 percent in 2001, but remains more rapid than China's, which decelerated from 1.4 percent to 0.7 percent during the same period. This presents a long-term opportunity for India but also a short-term problem because it places enormous demands on the government to find productive jobs for new entrants in the labor force—a huge challenge because services employment is expected to add only a few million jobs (albeit high-quality ones), and India's labor force numbers around 500 million, most of whom are still working on farms. Many of these people were probably among the voters dissatisfied that India's services-led prosperity was too narrowly based.

The size of India's labor force is likely to exceed that of China three decades from now, creating continued pressures on the economy to create jobs. But the large pool of labor—providing the country improves and broadens the benefits of education and training and can create productive jobs—could be turned into an advantage in coming decades as the working-age population increases. In contrast, those of other large emerging economies are projected to decelerate.

India is behind China in capital accumulation—but poised to grow stronger in that area. China has been promoting capital accumulation for two decades by actively soliciting direct investment from abroad and by encouraging a high rate of saving at home. There is also a considerable gap in cumulative foreign direct investment—5 percent of GDP for India, compared to 36 percent for China. However, since the early 1990s India—at the federal and state level—has been working to obtain more direct investment from abroad, and its savings rate also has been rising (now 24 percent of GDP), suggesting a potential for a higher rate of investment in coming years.

Robust investment growth will depend in part on whether India's fiscal deficit can be held in check, avoiding government deficits crowding out private investment. The government, faced with enormous demands to increase spending on the nation's poor, will face a big challenge in reducing budget deficits. The departed BJP government also sought to boost investment by improving the tax treatment of capital gains for new equity investments and infrastructure investment (such as electric power). Now 51 percent foreign ownership is permitted in important sectors of the economy, and the country has established export-processing zones in which 100 percent foreign ownership is permitted.

India continues to be burdened by what some investors and many Indians complain is excessive bureaucracy and in some areas is not as open an economy as China. While India has lowered tariffs considerably over the years—from 87 percent on the average in 1991 to 30 percent in early 2004—they are still three times the level of China (although in China administrative measures occasionally negate the benefits of lower tariffs for certain imports). And agriculture remains a vital aspect of the economy, employing around 65 percent of the workforce but turning out only a quarter of the GDP. A poor monsoon widens the budget deficit and weakens demand for other goods and services.

One important difference with China is the sectoral composition of the economy. India (as discussed above) has built up a large service sector. Offshore sourcing, largely from the U.S., has accelerated this trend, but the services boom had taken hold well before "offshoring" became so popular. India's share of world software and IT-related services has increased from 5 percent in 1997 to more than 20 percent. Services account for around 56 percent of the Indian economy, and 32 percent of exports, with manufacturing accounting for just over 22 percent of GDP and 49 percent of exports. By comparison, manufacturing accounts for about 40 percent of China's GDP, and 86 percent of its exports, while services account for about 30 percent of GDP and only 10 percent of its exports.

Several factors stand out when looking at future prospects for Indian growth. The number of Indians with high purchasing power—generally the portion of the country involved in business—has reached 60 million (over 6 percent of the population) representing 14 million households. The number of middle-class Indians—many involved in trade or prosperous commercial agricultural production—is nearly 300 million. These families are potential buyers of large amounts of middle to high-end consumer products such as refrigerators, televisions, and other consumer durables. The urban middle class is growing rapidly as the boom in high-technology centers creates demand for construction workers, building supplies, truck drivers, commercial food vendors, and domestic help.

Several sectors in India present possibilities for robust growth. The auto market is growing by over 20 percent annually. That growth is likely to continue as more middle–class consumers have the resources to buy cars. Production of industrial machinery for domestic use and export will increasingly take advantage of India's skilled workforce. Demand for basic materials is virtually certain to grow in India, as in China, to feed a growing economy (although the role of manufacturing in China's economy relative to that of India means that China's appetite for such goods will be a lot greater). The growth potential for telecommunications in India is high, for reasons discussed above, and there is likely to be a good opportunity for foreign firms to participate in this sector if the new government maintains the past government's relaxation of limits on foreign investment (maximum foreign investment permitted was raised to 75 percent from 49 percent). Mobile phones should experience particularly rapid growth in a country that lacks a large infrastructure of landlines. Cellular subscriptions have risen dramatically; falling prices for calls and growing numbers of middle-income consumers should lead to further rapid increases.

Given the growth potential in these areas, there has been increasing interest among foreign companies in establishing operations in India or entering into partnerships with Indian companies. In business process outsourcing, investment growth is robust. Investor confidence was enhanced by the last government's announcement allaying concerns about potential adverse tax treatment of IT companies. Annual growth in that sector has risen by 43 percent over the past five years and exports continue to rise rapidly.

What has added to the attractiveness of India as a place for investment is the demonstrated success of private sector innovation and entrepreneurship. Indian companies have become world-class competitors in software, business services, various other types of information technology, and pharmaceuticals. And Indian companies have been actively buying companies abroad.

India's economic future depends in substantial measure on the maintenance of stable political conditions and continued government implementation of policy reforms that give greater latitude to the private sector and growing opportunity to the country's rising group of entrepreneurs while also reducing income disparities and enhancing upward mobility. A welcoming environment for much-needed infrastructure investment—which the Congress government is fostering—is especially vital.

As with many emerging economies, institutions are important to success. Efficient and flexible institutions improve the confidence of individuals to invest and save, strengthening economic efficiency. India has a very robust institutional structure. It has a strong and credible legal system, well-regarded property rights protections, relatively transparent markets, and effective financial regulation and supervision. The new government remains committed to the positive evolution of such institutions; it is pledged to further enhance market-based policies.

Additional reforms will be important to sustained economic progress. Despite concerns of some members of the Indian business community, the previous government continued to protect small-scale manufacturing industries by limiting competition. This policy limited the expansion of large companies in these sectors. That in turn reduced consolidation and limited India's ability to compete internationally in some areas of manufacturing. And labor laws remain a constraint for business, because they still limit the ability of companies to lay off workers without agreement by state governments and trade unions—and therefore discourage hiring. India, like China, is concerned that too much labor market deregulation could lead to sharp increases in urban unemployment due to business shutdowns or layoffs. So movement in such areas will need to take into account social as well as economic requirements.

The country needs to expand basic education considerably (44 percent of Indians over the age of 15 have no schooling, 18 percent in China) if it is to realize its growth potential. That is a high priority of the government—as a way to broaden the benefits of prosperity, especially to rural India. It also aims to improve the country's weak physical infrastructure (which in recent years received only 7 percent of public spending as opposed to 36 percent for China).

Reforms in the financial sector illustrate the flexibility that India is capable of and the fact that the country is supportive of substantial reform efforts. They also demonstrate the benefits that can result from extensive reforms. Among the most significant reforms in this area has been the broadened access to bank credit. The former government rescued a number of public-sector financial institutions and development banks that had faced bankruptcy, improving confidence in the financial system. New laws on the recovery of bad loans have improved the liquidity of the banking system and enabled consumer credit to grow. India's financial markets now are relatively transparent, large numbers of companies can raise capital within the country, and the nation's banks have only a modest number of nonperforming loans.

The once large public sector now includes only three sectors (defense, railways, and nuclear power). Foreign investment limits in various sectors (such as banking, insurance, and telecom) have been liberalized. And significant improvements in corporate governance have taken place (such as requirements for a minimum number of independent directors, stricter accounting standards, and an audit committee).

Looking to the future, there have been numerous conversations at the official and private sector levels about ways in which China and India could collaborate on economic issues. Bilateral trade is expected to increase significantly. The two governments and the private sector are examining new areas for economic interaction. Potential areas include IT services/production, pharmaceuticals, telecom, and energy/environmental technology.

The fundamental commitment of the Congress-led government to continued reform, the talented experience of the nation's leadership, the entrepreneurial drive of its people, and the inherent strengths of the Indian economy are likely to remain important factors in India's very positive growth outlook. India is likely to be one of the world's great economic success stories in the first half of the twenty-first century, becoming the third largest economy in the world by 2050 with a GDP of between $25 and $30 trillion (in 2003 U.S. dollars).

Growth in Spurts

Asia is likely to experience a period of robust growth in this decade (more than 7 percent from Asia outside Japan)—but not necessarily a smooth one. Uncertainties such as the sustainability of China's red-hot growth and the impact on the region of higher interest rates in the U.S. or a fall in the dollar—combined with the less conventional risks of health epidemics, terrorism, and incidents of political unrest—are potential clouds on an otherwise bright horizon.

The key to understanding Asia is to realize that its economies are moving at a very rapid pace—with the aim of producing higher value-added, competitive goods and services and creating jobs for its rising and aspiring population. Mistakes are likely to be made. Booms and boomlets could occur from time to time—as in the past—only to be followed by busts and failures. Stock and bond markets are likely to fluctuate erratically on occasion, and the less transparent, the greater the risk. Policies will also have to adjust to address the continued poverty and income disparities in much of the region. Nonetheless, the overall direction in much of Asia appears to be very positive. Asia has the potential to deliver good returns to investors with the wisdom to understand the region's economies, cultures, and businesses—and the patience to see the process through.

Afterword

It is customary to believe that we must learn the first principles of economics before we can start to understand or interpret the real economy and the headlines that assail us each day. As someone who has toiled in both pastures—theory as well as practice—I offer the notion that perhaps we should reverse the order in which we approach the subject. We will achieve a deeper comprehension of theory, and a richer ability to apply its teachings, if we have first studied a wide range of practical analysis of real events.

As a result, I found reading this book to be a wonderfully revitalizing experience. The contributing authors restore one's faith in economics as a discipline and as a means to understanding the world around us.

If you pick up any economic journal today, or if you venture into the textbooks, you will find so many mathematical equations scattered throughout that you would think economics is a science. It is not a science, with principles immutable for all times and all places. Economics at its best is a disciplined thought process for analyzing the capitalist system and human interactions in the search for wealth and security. At its roots, economics is about how and why our society has changed and developed over time. Even deeper, economics is about risk and return.

These are the themes that infuse the contributions to this book. Although the keen insights, original diagnoses, and rare lucidity of the contributors to this volume inform us about the serious problems we face in today's world, that is by no means all they have to tell us. They have shaped their presentations around the primary elements of economic analysis: supply, demand, expectations, the critical role of real investment, foreign trade and finance, monetary theory and policy, and the interplay between the private and the public sectors.

The result is economics at its best—rich in description, searching in analysis, provocative in argument, profound in generalization, and always focused on the key issues. I am much the wiser for having read it.

—PETER L. BERNSTEIN

Index

About Bloomberg

Bloomberg L.P., founded in 1981, is a global information services, news, and media company. Headquartered in New York, the company has sales and news operations worldwide.

Bloomberg, serving customers on six continents, holds a unique position within the financial services industry by providing an unparalleled range of features in a single package known as the BLOOMBERG PROFESSIONAL® service. By addressing the demand for investment performance and efficiency through an exceptional combination of information, analytic, electronic trading, and Straight Through Processing tools, Bloomberg has built a worldwide customer base of corporations, issuers, financial intermediaries, and institutional investors.

BLOOMBERG NEWS®, founded in 1990, provides stories and columns on business, general news, politics, and sports to leading newspapers and magazines throughout the world. BLOOMBERG TELEVISION®, a 24-hour business and financial news network, is produced and distributed globally in seven languages. BLOOMBERG RADIO℠ is an international radio network anchored by flagship station BLOOMBERG® 1130 (WBBR-AM) in New York.

In addition to the BLOOMBERG PRESS® line of books, Bloomberg publishes *BLOOMBERG MARKETS®* and *BLOOMBERG WEALTH MANAGER®* magazines. To learn more about Bloomberg, call a sales representative at:

About the Editor

Thomas R. Keene, CFA is an editor-at-large who provides economic and investment perspective to all of Bloomberg News—print, radio, and television.

He writes the chart-of-the-day story that is available on the Bloomberg professional service and he also presents it on Bloomberg Television. Mr. Keene is the host of the weekday program *Bloomberg on the Economy* on Bloomberg Radio. Previously he worked in the investment industry for sixteen years.

Mr. Keene studied engineering at the University of Colorado and graduated from the Rochester Institute of Technology with a bachelor's degree in biology, cum laude.

He is a member of the National Association for Business Economics and the CFA Institute, the organization that administers the Chartered Financial Analyst program. He is working toward a further degree from the London School of Economics. He lives in Boston and in New York City.

A Note on the Type

The text of this book is set in Bembo, in a modern version modeled on a typeface first cut by the goldsmith Francesco Griffo in 1495 for the Venetian printer Aldus Manutius and used for Pietro Bembo's small treatise, *De Aetna*. The Bembo face significantly influenced European typeface design and was a forerunner of faces designed by Claude Garamond (1480–1561). The present-day form was designed for the Monotype Corporation, London, in 1929.